T0135997

Brain and Cognitive Intelligence Control in Robotics

Editor

Bin Wei
Department of Computer Science
Algoma University
Sault Ste Marie, Ontario, Canada

CRC Press
Taylor & Francis Group
Boca Raton   London   New York

CRC Press is an imprint of the
Taylor & Francis Group, an **informa** business
A SCIENCE PUBLISHERS BOOK

First edition published 2022
by CRC Press
6000 Broken Sound Parkway NW, Suite 300, Boca Raton, FL 33487-2742

and by CRC Press
4 Park Square, Milton Park, Abingdon, Oxon, OX14 4RN

© 2022 Taylor & Francis Group, LLC

CRC Press is an imprint of Taylor & Francis Group, LLC

Library of Congress Cataloging-in-Publication Data (applied for)

ISBN: 978-0-367-50552-3 (hbk)
ISBN: 978-0-367-50553-0 (pbk)
ISBN: 978-1-003-05031-5 (ebk)

DOI: 10.1201/9781003050315

Typeset in Times New Roman
by Radiant Productions

Preface

This book, Brain and Cognitive Intelligence: Control in Robotics, will focus on the topics of cognitive and learning based intelligent control and its applications in robotics.

I would like to thank all the authors for their contributions to the book. I am also grateful to the publisher for supporting this project. I hope the readers find this book informative and useful.

This book consists of 5 chapters. Chapter 1 presents a real-time kinodynamic motion planning technique for linear systems with completely unknown dynamics in environments with unpredictable obstacles. The methodology incorporates: (i) a sampling-based algorithm for path planning and fast re-planning; and (ii) continuous-time Q-learning for the solution of finite-horizon optimal control problems in real-time. Chapter 2 present a semi-automated seed delivery and tracking scheme for precise seed placement, to allow for more conformal treatment of localized prostate cancer. Standard seed-carrying needles are connected to a hand-held apparatus that the surgeon holds to steer the needle automatically as the surgeon manually inserts it. The apparatus is equipped with force sensors that are used to estimate the tissue properties during insertion. A needle steering controller employs ultrasound images of the needle in tissue to calculate the optimal steering manoeuvres online. A method is derived to track the position of the implanted seeds after the needle is withdrawn, allowing the surgeon to monitor the implant quality online. Chapter 3 presents team cognition assessment: from concept to practice. It talked about the concepts of team and team composition, team cognition, and team cognition assessment, also the pieces of evidence presented in this chapter were collected from healthcare, results can be applied to other industries that are heavily dependent on team works. As new tracking technologies will be available for studying team collaboration, it is anticipated that more behavioral pieces of evidence will be produced to the goal of assessing team performance and cognition with objective methods. In Chapter 4, it is found that people's decisions are based on bounded rationality: since we cannot compute an optimal solution for all possible situations, we divide situations into groups and come up with a solution appropriate for each group. What is optimal here is the division into groups. It is therefore desirable to implement a similar algorithm for robots. To help with such algorithms, we provide techniques that help optimally divide situations into groups. Chapter 5 presented the neural-computer interfaces, Embodied cultured networks, cybernetic organisms, bioengineering mini brain hybrots, applications for minimal cognition research and ethical and legal considerations.

Finally, the editor would like to acknowledge all the friends and colleagues who have contributed to this book.

Bin Wei

September 24, 2021   Sault Ste Marie, Ontario, Canada

Contents

Chapter 1

RRT-QX

Real-Time Kinodynamic Motion Planning in Dynamic Environments with Continuous-Time Reinforcement Learning

George P Kontoudis,[a],* *Kyriakos G Vamvoudakis*[b] and *Zirui Xu*[c]

1. Introduction

Substantial improvements in artificial intelligence, computing resources, and software tools have enabled tremendous capabilities to mobile robots and autonomous systems. The problem of navigation is a core topic in robotics and autonomous vehicles, as the majority of robotic applications require safe path planning and obstacle avoidance (Yang et al., 2018). Ideally, a solution to this problem considers collision-free navigation in dynamic environments, computationally affordable algorithms for real-time implementation, and optimal control strategies. Such a challenging problem should be addressed in the continuous-time domain, as naive discretization of the system dynamics and the policy space, disregards critical information and leads to discretization errors (Lillicrap et al., 2015). In addition, dynamic environments impose time constraints to the motion planning problem, because collision-free navigation is only ensured for limited time frames (Nägeli et al., 2017). The latter necessitates a finite-horizon formulation to the optimal control problem. Moreover, system modeling is a challenging task with

[a] Maryland Robotics Center, University of Maryland, College Park, MD, USA.

[b] Guggenheim Sch. of Aerospace Eng., Georgia Institute of Technology, Atlanta, GA, USA.

[c] Department of Aerospace Engineering, University of Michigan, Ann Arbor, MI, USA.

* Corresponding author: gpkont@vt.edu

inevitable model simplifications and inaccuracies (Berkenkamp and Schoellig, 2015). Thus, a combination of optimal and adaptive control is needed. Finally, even if the system dynamics are assumed to be known, the finite-horizon optimal control problem requires extensive offline computations to solve the Hamilton-Jacobi-Bellman equation (Lewis et al., 2012).

Our aim in this work is to present a real-time kinodynamic motion planning technique for dynamic environments with unpredictably appearing obstacles. We address the finite-horizon optimal control problem with completely unknown system dynamics. The unknown model is considered to be continuous-time linear time-invariant.

Motion planning in high-dimensions has been addressed with probabilistic road-maps (PRM) (Kavraki et al., 1996) and rapidly-exploring random trees (RRT) (Kuffner and LaValle, 2000; LaValle, 1998). These algorithms are probabilistically complete, but not optimal. The work of Karaman and Frazzoli (2011) proposed a variation of RRT based on rewiring, namely RRT*. The latter was proved to be probabilistically complete and asymptotically optimal. These methods do not incorporate realistic system dynamics and instead use simple dynamics.

The problem of kinodynamic motion planning is introduced in (Donald et al., 1993). Kinodynamic RRT (LaValle and Kuffner, 2001) employs the dynamical model of the system, but the proposed control strategy is selected either randomly or by testing multiple controls and selecting the best. LQR-trees (Tedrake et al., 2010) is a feedback motion planning algorithm that utilizes optimization tools. This method requires significant computations to solve the Riccati equation. A combination of linear quadratic regulator (LQR) and RRT* is proposed in (Perez et al., 2010). In particular, the authors formulate a free-final-state, infinite-horizon optimal control problem with minimum energy cost and a heuristic extension of the RRT*. This algorithm incorporates the system dynamics and enforces extensive offline computations. Kinodynamic RRT* (Webb and Van Den Berg, 2013) is an asymptotically optimal motion planner for known linear time-invariant systems. The authors formulate a finite-horizon optimal control problem of fixed-final-state and free-final-time with minimum fuel-time performance. However, kinodynamic RRT* yields an open-loop controller and the computation of the continuous reachability Gramian requires significant offline computation. The authors in (Li et al., 2018) proposed a near optimal kinodynamic motion planning technique, that is named NoD-RRT. The methodology utilizes neural network approximation and RRT. NoD-RRT achieved reduced computational complexity and enhanced performance for nonlinear systems, comparing to RRT and kinodynamic RRT*. Yet, their framework is model-based and requires offline computations. In (Kontoudis and Vamvoudakis, 2019a), the authors presented RRT-Q*, an online, model-free kinodynamic motion planning framework which computes approximately optimal control policies for motion planning in static environments. RRT-Q* combines continuous-time Q-learning, RRT*, and local replanning in relatively small spaces. The latter has been robustified in (Kontoudis and Vamvoudakis, 2019b). In (Chiang et al., 2019), the authors combined reinforcement learning (RL) and RRT for kinodynamic motion planning. RL is used to learn obstacle avoiding policies and supervised learning to predict the time to reach a state and guide the growth of the tree. However, this algorithm requires significant offline computations. All the aforementioned motion planning techniques can only deal with static environments.

Randomized kinodynamic motion planning in dynamic environments is introduced in (Hsu et al., 2002). In (Bruce and Veloso, 2002), the execution-extended RRT is presented for real-time replanning. This algorithm stores selective nodes in a waypoint cache and performs an iterative, adaptive cost search on the forward tree, towards efficient replanning in dynamic domains. Dynamic RRT (Ferguson et al., 2006) trims the invalid leafs of the tree when a collision occurs due to a new obstacle configuration and grows the rest tree from the goal to the current configuration of the robot. The authors in (Otte and Frazzoli, 2014, 2016) present the RRT^X, an asymptotically optimal motion planning algorithm for both static and dynamic environments. RRT^X has the ability to perform quick online replanning. In (Allen and Pavone, 2019), the authors proposed an online kinodynamic motion planning algorithm which was experimentally validated in dynamic indoor environments. The technique requires the model of the system and its efficiency depends on the offline training of reachability sets. Kontoudis et al. (2020) combined event-triggered Q-learning and RRT^X to address the kinodynamic motion planning problem in dynamic domains. In addition, cognitive hierarchy along with RRT^X and Q-learning has been used to multi-robot motion planning in human-crowded environments Netter et al. (2021).

Optimal control (Lewis et al., 2012) can be efficiently merged with adaptive control (Ioannou and Sun, 2012) by employing principles of reinforcement learning (Sutton and Barto, 2018), and approximate dynamic programming (Busoniu et al., 2010; Lewis et al., 2012; Powell, 2007; Vrabie et al., 2013). In (Watkins and Dayan, 1992), a solution to Markovian systems was proposed with the use of discrete-time Q-learning. The authors in (Mehta and Meyn, 2009), presented a connection of Q-learning with nonlinear control based on the observation that the Q-function is related to the Hamiltonian. A solution to the model-free, infinite horizon optimal control problem for continuous-time linear time-invariant systems is presented in (Vamvoudakis, 2017).

The remainder of this paper is organized as follows. In Section 2 we formulate the problem, Section 3 discusses the optimal control problem, Section 4 provides a model-free formulation based on Q-learning, and Section 5 presents the structure of RRT-Q*. Section 6 illustartes the efficiency of our method through simulations and Section 7 concludes the chapter.

The notation here is standard. The set of real numbers is denoted $\mathbb{R}$, the set of all positive real numbers $\mathbb{R}^+$, the set of $n \times m$ real matrices $\mathbb{R}^{n \times m}$, and the set of natural numbers $\mathbb{N}$. The notation $(\cdot)^\mathsf{T}$ and $(\cdot)^{-1}$ denote the transpose and inverse operator respectively. The superscript $\star$ denotes the optimal solutions of a minimization problem. The notations $\underline{\lambda}(A)$ and $\overline{\lambda}(A)$ denote the minimum and maximum eigenvalues of the matrix A respectively. We denote $\mathrm{vech}(A)$, $\mathrm{vec}(A)$, and $\mathrm{mat}(A)$ the half-vectorization, vectorization, and matrization of a matrix A respectively. The Minkowski sum of two sets is denoted $\oplus$. A positive and semi-positive definite A matrix is denoted by $A \succ 0$ and $A \succeq 0$ respectively. The notation $U \otimes V$ denotes the Kronecker product of two vectors and $\| \cdot \|$ denotes the L_2 norm.

2. Problem Formulation

Let a linear time-invariant continuous-time system,

$$\dot{x}(t) = Ax(t) + Bu(t), \ x(0) = x_0, \ t \geq 0,$$

where $x(t) \in \mathcal{X} \subseteq \mathbb{R}^n$ is the state vector, $u(t) \in \mathbb{R}^m$ is the control input, and $A \in \mathbb{R}^{n \times n}$, $B \in \mathbb{R}^{n \times m}$ are the unknown plant and input matrices respectively. To drive our system from an initial state x_0 to a final state $x(T) = x_\mathrm{r}$, we define the difference between the state $x(t)$ and the state x_r, as our new state $\bar{x}(t) := x(t) - x_\mathrm{r}$. The final time is denoted by $T \in \mathbb{R}^+$. The new system yields,

$$\dot{\bar{x}}(t) = A\bar{x}(t) + Bu(t), \ \bar{x}_0 = x_0 - x_{\mathrm{r}}, \ t \geq 0. \tag{1}$$

Consider an energy cost function,

$$J(\bar{x}; u; t_0, T) = \phi(T) + \frac{1}{2} \int_{t_0}^{T} (\bar{x}^\mathsf{T} M \bar{x} + u^\mathsf{T} R u) \, \mathrm{d}\tau, \ \forall t_0, \tag{2}$$

where $\phi(T) = \frac{1}{2}\bar{x}^\mathsf{T}(T)P(T)\bar{x}(T)$ is the terminal cost with $P(T) = P_{\mathrm{T}} \in \mathbb{R}^{n \times n} \succ 0$ the final Riccati matrix, $M \in \mathbb{R}^{n \times n} \succeq 0$ and $R \in \mathbb{R}^{m \times m} \succ 0$ user defined matrices that penalize the states and the control input respectively.

Assumption 1 *The unknown pairs (A,B) and ($\sqrt{M}$, A) are controllable and detectable respectively.*

We are interested in finding an optimal control $u^\star$ such that it satisfies $J(\bar{x}; u^\star; t_0, T) \leq J(\bar{x}; u; t_0, T)$, $\forall \bar{x}, u$, which can be described by the minimization problem $J(\bar{x}; u^\star; t_0, T) = \min_u J(\bar{x}; u; t_0, T)$ subject to (1). In other words, we want to obtain the optimal value function $V^\star$ that is defined by,

$$V^\star(\bar{x}; t_0, T) := \min_u \left(\phi(T) + \frac{1}{2} \int_{t_0}^{T} (\bar{x}^\mathsf{T} M \bar{x} + u^\mathsf{T} R u) \, \mathrm{d}\tau \right), \tag{3}$$

but without any information about the system dynamics.

Consider the known obstacle closed space $\mathcal{X}_{\mathrm{obs}} \subset \mathcal{X}$. For multiple obstacles, the obstacle space is defined $\mathcal{X}_{\mathrm{obs}} := \bigcup_{l=1}^{N_o} \mathcal{X}_{\mathrm{obs},l}$, where $N_o \in \mathbb{N}$ is the total number of obstacles. Thus, the free space is an open space $\mathcal{X}_{\mathrm{free}} = (\mathcal{X}_{\mathrm{obs}})^\complement = \mathcal{X} \backslash \mathcal{X}_{\mathrm{obs}}$. In dynamic environments, the obstacle space $\mathcal{X}_{\mathrm{obs}}$ and the free space $\mathcal{X}_{\mathrm{free}}$ evolve in time. We define the unpredictable variation of the obstacle space as $\Delta \mathcal{X}_{\mathrm{obs}} := f(\mathcal{X}_{\mathrm{obs}}; t)$, where $f(\cdot)$ is unknown, and $\Delta \mathcal{X}_{\mathrm{obs}} = \{\varnothing\}$ indicates no obstacle changes in the environment. We use RRT$^\mathrm{X}$ for path planning. RRT$^\mathrm{X}$ constructs a graph $G = (V, E)$, where V is the set of nodes and E is the set of edges. As a slight abuse of notation, we will refer to nodes $v \in V$ as states $x \in \mathcal{X}$. The planner provides an optimal sub-tree that contains the planned path $\pi(x_{0,k}, x_{\mathrm{r},k}; t) \in \mathbb{R}^{2(K \times n)}$, where $k = 1, \ldots, K$, $K \in \mathbb{N}$ is the number of boundary value problems (BVPs). Each BVP is described by the initial and desired state $(x_{0,k}, x_{\mathrm{r},k})$. Since the obstacle space $\mathcal{X}_{\mathrm{obs}}$ evolves in time, π is also a function of time, and thus K also change accordingly.

We seek to drive the system to a desired state, without any knowledge of the system dynamics. For the k-th BVP, let us define the initial distance as the distance from the initial state $x_{0,k}$ to the desired state $x_{r,k}$,

$$D_0(\bar{x}_{0,k}) := \|x_{0,k} - x_{r,k}\| = \|\bar{x}_{0,k}\|, \quad \forall \bar{x}_0 \in \mathbb{R}^n, \quad (4)$$

and the relative distance to $x_{r,k}$,

$$D(\bar{x}) := \|x - x_{r,k}\| = \|\bar{x}\|, \quad \forall \bar{x} \in \mathbb{R}^n. \quad (5)$$

Since we address the free-final state finite-horizon optimal control problem, the controller at final time T approximates the desired state x_r, i.e., $x(T)$ converges to a close neighborhood around x_r (Bryson, 1975; Lewis et al., 2012). In order to reduce the navigation time, we assume that the desired state is reached when the system enters the close neighborhood of the desired final state x_r. That is to say, when $D(\bar{x}) \leq \rho D_0(\bar{x}_{0,k})$, where ρ is the user-defined admissible window, the robot is considered to have reached the desired state $x_{r,k}$. Subsequently, the system proceeds to the next $(k+1)$-th problem.

Moreover, since the system dynamics are unknown, when RRT^X calculates the collision-free path π, it can only adopt straight lines as edges in E. However, the actual trajectory of the robot is curved due to the kinodynamic constraints (1) and the optimal performance (2). Thus, the actual trajectory deviates from the nominal trajectory provided by the RRT^X, and collisions may occur in near-to-obstacle areas. To address this issue, we introduce an obstacle augmentation strategy. More specifically, the algorithm computes at every time instance the kinodynamic distance,

$$D_{\text{rob}}(\bar{x}) := \frac{|\bar{x}_{0,k} \times \bar{x}|}{D_{0,k}}, \quad (6)$$

which represents the present current deviation of the robot's current position from the nominal trajectory, i.e., corresponding straight path determined by $(x_{0,k}, x_{r,k})$. Then, an augmented obstacle space $\mathcal{X}_{\text{obs}}^{\text{aug}}$ is obtained from,

$$\mathcal{X}_{\text{obs}}^{\text{aug}} := \mathcal{X}_{\text{obs}} \oplus \mathcal{X}_{\text{kin}}, \quad (7)$$

where $\mathcal{X}_{\text{kin}}$ is the space of a compact set bounded by a circle with radius equal to the maximum kinodynamic distance $D_{\text{rob}}^{\text{kin}}$. When the maximum kinodynamic distance is updated, RRT^X provides a new path based on the the newest augmented obstacle space (7).

3. Finite-Horizon Boundary Value Problem

Let us define the Hamiltonian with respect to (1) and (3) as,

$$\mathcal{H}(\bar{x}; u; \lambda) := \frac{1}{2}(\bar{x}^\mathsf{T} M \bar{x} + u^\mathsf{T} R u) + \lambda^\mathsf{T}(A\bar{x} + Bu), \ \forall \bar{x}, u, \lambda.$$

In order to solve the finite-horizon optimal control problem (3), we use the sweep method (Bryson, 1975) and we set $\lambda = \frac{\partial V^\star}{\partial \bar{x}}$. Thus, the Hamilton-Jacobi-Bellman (HJB) equation yields,

$$-\frac{\partial V^\star}{\partial t} = \frac{1}{2}(\bar{x}^\mathsf{T} M \bar{x} + u^\mathsf{T} R u) + \frac{\partial V^\star}{\partial \bar{x}}^\mathsf{T}(A\bar{x} + Bu), \ \forall \bar{x}.$$

Since our system (1) is linear, we write the value function in a quadratic form as,

$$V^\star(\bar{x}; t) = \frac{1}{2}\bar{x}^\mathsf{T} P(t)\bar{x}, \ \forall \bar{x}, t \geq t_0, \tag{8}$$

where $P(t) \in \mathbb{R}^{n \times n} \succ 0$ is the Riccati matrix that solves the differential Riccati equation,

$$-\dot{P}(t) = M + P(t)A + A^\mathsf{T} P(t) - P(t)BR^{-1}B^\mathsf{T} P(t). \tag{9}$$

Hence, the optimal control gets the form of,

$$u^\star(\bar{x}; t) = -R^{-1}B^\mathsf{T} P(t)\bar{x}, \ \forall \bar{x}, t. \tag{10}$$

Theorem 3.1
Suppose that there exists a $P(t) \succ 0$ that satisfies the Riccati equation (9) with a final condition given by P_T, and the control obtained by,

$$u(\bar{x}; t) = -R^{-1}B^\mathsf{T} P(t)\bar{x}. \tag{11}$$

Then, the control input (11) minimizes the cost given in (3), and the origin is a globally uniformly asymptotically stable equilibrium point of the closed-loop system.

Proof The proof follows from (Kontoudis and Vamvoudakis, 2019a).

□

4. Model-Free Formulation

Let us now define the following Q-function as,

$$\mathcal{Q}(\bar{x}; u; t) := V^\star(\bar{x}; t) + \mathcal{H}(\bar{x}; u; \frac{\partial V^\star}{\partial t}, \frac{\partial V^\star}{\partial \bar{x}})$$

$$= V^\star(\bar{x}; t) + \frac{1}{2}\bar{x}^\mathsf{T} M\bar{x} + \frac{1}{2}u^\mathsf{T} Ru + \bar{x}^\mathsf{T} P(t)(A\bar{x} + Bu) + \frac{1}{2}\bar{x}^\mathsf{T} \dot{P}(t)\bar{x},$$

$$(12)$$

where $\mathcal{Q}(\bar{x}; u; t) \in \mathbb{R}$ is an action-dependent value.

Next, we define the augmented state $U := [\bar{x}^\mathsf{T}\ u^\mathsf{T}]^\mathsf{T} \in \mathbb{R}^{(n+m)}$ to express the Q-function (12) in a compact form as,

$$\mathcal{Q}(\bar{x}; u; t) = \frac{1}{2}U^\mathsf{T} \begin{bmatrix} Q_{\mathrm{xx}}(t) & Q_{\mathrm{xu}}(t) \\ Q_{\mathrm{ux}}(t) & Q_{\mathrm{uu}} \end{bmatrix} U =: \frac{1}{2}U^\mathsf{T} \bar{\mathcal{Q}}(t)U, \qquad (13)$$

where $Q_{\mathrm{xx}}(t) = \dot{P}(t) + P(t) + M + P(t)A + A^\mathsf{T} P(t)$, $Q_{\mathrm{xu}}(t) = Q_{\mathrm{ux}}^\mathsf{T}(t) = P(t)B$, and $Q_{\mathrm{uu}} = R$, with $\mathcal{Q} : \mathbb{R}^{n+m} \times \mathbb{R}^{(n+m)\times(n+m)} \to \mathbb{R}$. Using the stationarity condition $\partial \mathcal{Q}(\bar{x}; u; t)/\partial u = 0$, we find a model-free expression of the optimal control $u^\star$ (10) as,

$$u^\star(\bar{x}; t) = \arg\min_u \mathcal{Q}(\bar{x}; u; t) = -Q_{\mathrm{uu}}^{-1} Q_{\mathrm{ux}}(t)\bar{x}. \qquad (14)$$

Lemma 4.1
The value of the minimization $\mathcal{Q}^\star(\bar{x}; u^\star; t) := \min_u \mathcal{Q}(\bar{x}; u; t)$ is the same with the optimal value $V^\star$ in (8) of the minimization problem (3), where $P(t) \succ 0$ is the Riccati matrix found from (9).

Proof The proof follows from (Kontoudis and Vamvoudakis, 2019a).

□

4.1 Actor/Critic Network

A critic approximator is designed to approximate the Q-function in (13) as,

$$\mathcal{Q}^\star(\bar{x}; u^\star; t) = \frac{1}{2}U^\mathsf{T} \begin{bmatrix} Q_{\mathrm{xx}}(t) & Q_{\mathrm{xu}}(t) \\ Q_{\mathrm{ux}}(t) & Q_{\mathrm{uu}} \end{bmatrix} U := \frac{1}{2}\mathrm{vech}(\bar{\mathcal{Q}}(t))^\mathsf{T}(U \otimes U),$$

where $\mathrm{vech}(\bar{\mathcal{Q}}(t)) \in \mathbb{R}^{\frac{(n+m)(n+m+1)}{2}}$. The half-vectorization operation exploits the symmetric properties of the $\bar{\mathcal{Q}}$ matrix to reduce the computations. Then, by setting $\nu(t)^\mathsf{T} W_\mathrm{c} := 1/2\mathrm{vech}(\bar{\mathcal{Q}}(t))$ we obtain,

$$\mathcal{Q}^\star(\bar{x}; u^\star; t) = W_\mathrm{c}^\mathsf{T}\nu(t)(U \otimes U),$$

where $W_c \in \mathbb{R}^{\frac{(n+m)(n+m+1)}{2}}$ is the critic weight estimator vector, and $\nu(t) \in \mathbb{R}^{\frac{(n+m)(n+m+1)}{2} \times \frac{(n+m)(n+m+1)}{2}}$ is a radial basis function of appropriate dimensions that depends explicitly on time. Since the ideal weight estimates are unknown, we employ an adaptive estimation technique (Ioannou and Sun, 2012) to approximate the Q-function,

$$\hat{\mathcal{Q}}(\bar{x}; u; t) = \hat{W}_c^{\mathsf{T}} \nu(t)(U \otimes U), \tag{15}$$

where $\hat{W}_c \nu(t) := \frac{1}{2}\mathrm{vech}(\hat{\bar{\mathcal{Q}}}(t))$.

By using a similar way of thinking for the actor we assign $\mu(t)^{\mathsf{T}} W_a := -Q_{uu}^{-1} Q_{ux}(t)$ to write,

$$u^\star(\bar{x}; t) = W_a^{\mathsf{T}} \mu(t)\bar{x},$$

where $W_a \in \mathbb{R}^{n \times m}$ is the actor weight estimator vector, $\mu(t) \in \mathbb{R}^{n \times n}$ is a radial basis function of appropriate dimensions that depends explicitly on time. The actor by using current weight estimates yields,

$$\hat{u}(\bar{x}; t) = \hat{W}_a^{\mathsf{T}} \mu(t)\bar{x}. \tag{16}$$

Remark 1. The approximation errors of the critic and the actor approximators described in (15) and (16) respectively, vanish as the system (1) is linear. To this end, we exploit the whole space and not just a compact set. With this structure, the approximations will converge to the optimal policies, and hence the superscript $\star$, that denotes the ideal values of the adaptive weight estimation, render similarly with the optimal solutions.

Next, we adopt an integral reinforcement learning approach (Vrabie et al., 2013) that lets us express the Bellman equation as,

$$V^\star(\bar{x}(t); t) = V^\star(\bar{x}(t - \Delta t); t - \Delta t) - \frac{1}{2} \int_{t-\Delta t}^{t} (\bar{x}^{\mathsf{T}} M \bar{x} + u^{\star\mathsf{T}} R u^\star)\, \mathrm{d}\tau, \tag{17}$$

$$V^\star(\bar{x}(T); T) = \frac{1}{2}\bar{x}^{\mathsf{T}}(T)P(T)\bar{x}(T), \tag{18}$$

where $\Delta t \in \mathbb{R}^+$ is a small fixed value, i.e., resolution. By following Lemma 4.1, where we proved that $\mathcal{Q}^\star(\bar{x}; u^\star; t) = V^\star(\bar{x}; t)$, we can write (17) and (18) as,

$$\mathcal{Q}^\star(\bar{x}(t); u^\star(t); t) = \mathcal{Q}^\star(\bar{x}(t - \Delta t); u^\star(t - \Delta t); t - \Delta t) - \frac{1}{2} \int_{t-\Delta t}^{t} (\bar{x}^{\mathsf{T}} M \bar{x} + u^{\star\mathsf{T}} R u^\star)\, \mathrm{d}\tau,$$

$$\mathcal{Q}^\star(\bar{x}(T); T) = \frac{1}{2}\bar{x}^{\mathsf{T}}(T)P(T)\bar{x}(T).$$

Next, we define the errors $e_{c_1}, e_{c_2} \in \mathbb{R}$, that we seek to drive to zero by appropriately tuning the critic weights of (15). Define the first critic error e_{c_1} as,

$$
\begin{aligned}
e_{c_1} &:= \hat{Q}(\bar{x}(t); \hat{u}(t); t) - \hat{Q}(\bar{x}(t - \Delta t); \hat{u}(t - \Delta t); t - \Delta t) \\
&\quad + \frac{1}{2} \int_{t - \Delta t}^{t} (\bar{x}^{\mathsf{T}} M \bar{x} + \hat{u}^{\mathsf{T}} R \hat{u}) \, \mathrm{d}\tau \\
&= \hat{W}_c^{\mathsf{T}} \Big(\nu(t)(U(t) \otimes U(t)) - \nu(t - T)(U(t - \Delta t) \otimes U(t - \Delta t)) \Big) \\
&\quad + \frac{1}{2} \int_{t - \Delta t}^{t} (\bar{x}^{\mathsf{T}} M \bar{x} + \hat{u}^{\mathsf{T}} R \hat{u}) \, \mathrm{d}\tau,
\end{aligned}
\tag{19}
$$

Intrinsic dynamics are included in (19), which can be evaluated by taking the time derivative,

$$
\dot{p} = \bar{x}^{\mathsf{T}}(t) M \bar{x}(t) - \bar{x}^{\mathsf{T}}(t - \Delta t) M \bar{x}(t - \Delta t) + \hat{u}^{\mathsf{T}}(t) R \hat{u}(t) - \hat{u}^{\mathsf{T}}(t - \Delta t) R \hat{u}(t - \Delta t).
$$

The second critic error is defined by,

$$
e_{c_2} := \frac{1}{2} \bar{x}^{\mathsf{T}}(t) P(T) \bar{x}(t) - \hat{W}_c^{\mathsf{T}} \nu(t)(U(t) \otimes U(t)).
$$

The actor approximator error $e_a \in \mathbb{R}^m$ is defined by,

$$
e_a := \hat{W}_a^{\mathsf{T}} \mu(t) \bar{x} + \hat{Q}_{uu}^{-1} \hat{Q}_{ux}(t) \bar{x},
$$

where $\hat{Q}_{uu}, \hat{Q}_{ux}$ will be obtained from the critic weight matrix estimation $\hat{W}_c$. By employing adaptive control techniques (Ioannou and Sun, 2012), we formulate the squared-norm of errors as,

$$
K_1(\hat{W}_c, \hat{W}_c(T)) = \frac{1}{2} \|e_{c_1}\|^2 + \frac{1}{2} \|e_{c_2}\|^2,
\tag{20}
$$

$$
K_2(\hat{W}_a) = \frac{1}{2} \|e_a\|^2.
\tag{21}
$$

4.2 Learning Methodology

The weights of the critic estimation matrix are obtained by applying a normalized gradient descent algorithm in (20),

$$
\dot{\hat{W}}_c = -\alpha_c \frac{\partial K_1}{\partial \hat{W}_c} = -\alpha_c \left(\frac{1}{(1 + \sigma^{\mathsf{T}} \sigma)^2} \sigma e_{c_1} + \frac{1}{(1 + \sigma_f^{\mathsf{T}} \sigma_f)^2} \sigma_f e_{c_2} \right),
\tag{22}
$$

where $\sigma(t) := \nu(t)(U(t) \otimes U(t) - U(t - \Delta t) \otimes U(t - \Delta t))$, $\sigma_{\mathrm{f}}(t) = \nu(t)(U(t) \otimes U(t))$, and $\alpha_{\mathrm{c}} \in \mathbb{R}^+$ is a constant gain that specifies the convergence rate. The critic tuning (22) guarantees that as $e_{c_1} \to 0$ and $e_{c_2} \to 0$ then $\hat{W}_{\mathrm{c}} \to W_{\mathrm{c}}$ and $\hat{W}_{\mathrm{c}}(T) \to W_{\mathrm{c}}(T)$.

Similarly, the weights of the actor estimation matrix $\hat{W}_{\mathrm{a}}$ by applying a gradient descent algorithm in (21) yield,

$$\dot{\hat{W}}_{\mathrm{a}} = -\alpha_{\mathrm{a}} \frac{\partial K_2}{\partial \hat{W}_{\mathrm{a}}} = -\alpha_{\mathrm{a}} \bar{x} e_{\mathrm{a}}^{\mathsf{T}}, \tag{23}$$

where $\alpha_{\mathrm{a}} \in \mathbb{R}^+$ is a constant gain that specifies the convergence rate. The actor estimation algorithm (23) guarantees that as $e_{\mathrm{a}} \to 0$ then $\hat{W}_{\mathrm{a}} \to W_{\mathrm{a}}$.

For the theoretical analysis we introduce the weight estimation error for the critic $\tilde{W}_{\mathrm{c}} := W_{\mathrm{c}} - \hat{W}_{\mathrm{c}}$ and for the actor $\tilde{W}_{\mathrm{a}} := W_{\mathrm{a}} - \hat{W}_{\mathrm{a}}$, with $\tilde{W}_{\mathrm{c}} \in \mathbb{R}^{\frac{(n+m)(n+m+1)}{2}}$, $\tilde{W}_{\mathrm{a}} \in \mathbb{R}^{n \times m}$. The estimation error dynamics of the critic yields,

$$\dot{\tilde{W}}_{\mathrm{c}} = -\alpha_{\mathrm{c}} \frac{1}{(1 + \sigma^{\mathsf{T}} \sigma)^2} \sigma \sigma^{\mathsf{T}} \tilde{W}_{\mathrm{c}},$$

and the estimation error dynamics of the actor becomes,

$$\dot{\tilde{W}}_{\mathrm{a}} = -\alpha_{\mathrm{a}} \bar{x} \bar{x}^{\mathsf{T}} \mu(t)^{\mathsf{T}} \tilde{W}_{\mathrm{a}} - \alpha_{\mathrm{a}} \bar{x} \bar{x}^{\mathsf{T}} \frac{\mu(t) \tilde{Q}_{\mathrm{xu}} R^{-1}}{\|1 + \mu(t)^{\mathsf{T}} \mu(t)\|^2}, \tag{24}$$

where $\tilde{Q}_{\mathrm{xu}} := \mathrm{mat}(\tilde{W}_{\mathrm{c}}[\frac{n(n+1)}{2} + 1 : \frac{n(n+1)}{2} + nm])$.

Lemma 4.2
For any given control input $u(t) \in \mathcal{U}$ the estimation error dynamics of the critic (24) have an exponentially stable equilibrium point at the origin as follows,

$$\|\tilde{W}_{\mathrm{c}}\| \leq \|\tilde{W}_{\mathrm{c}}(t_0)\| \kappa_1 e^{-\kappa_2(t - t_0)},$$

where $\kappa_1, \kappa_2 \in \mathbb{R}^+$. In order to establish exponential stability, we require the signal $\Delta(t) := \frac{\sigma(t)}{1 + \sigma(t)^{\mathsf{T}} \sigma(t)}$ to be persistently exciting (PE) at $[t, t + T_{PE}]$, where $T_{PE} \in \mathbb{R}^+$ the excitation period, if there exists a $\beta \in \mathbb{R}^+$ such that $\beta I \leq \int_t^{t + T_{PE}} \Delta(\tau) \Delta^{\mathsf{T}}(\tau) d\tau$, where I is an identity matrix of appropriate dimensions.

Proof The proof follows from (Vamvoudakis, 2017). □

The main stability theorem for the Q-learning framework is provided below.

Theorem 4.3
Consider the linear time-invariant continuous-time system (1), *the critic, and the actor approximators given by* (15), *and* (16) *respectively. The weights of the critic, and the actor estimators are tuned by* (22), *and* (23) *respectively. The origin with state* $\psi = [\bar{x}^\mathsf{T} \ \tilde{W}_c^\mathsf{T} \ \tilde{W}_a^\mathsf{T}]^\mathsf{T}$ *is a globally uniformly asymptotically stable equilibrium point of the closed-loop system and for all initial conditions* $\psi(0)$, *given that the critic gain* α_c *is sufficiently larger than the actor gain* α_a *and the following inequality holds,*

$$0 < \alpha_a < \frac{2\underline{\lambda}(M + Q_{xu}R^{-1}Q_{xu}^\mathsf{T}) - \overline{\lambda}(Q_{xu}Q_{xu}^\mathsf{T})}{\delta\overline{\lambda}\left(\frac{\mu(t)R^{-1}}{\|1 + \mu(t)^\mathsf{T}\mu(t)\|^2}\right)}, \qquad (25)$$

with δ *a constant of unity order.*

Proof The proof follows from (Kontoudis and Vamvoudakis, 2019a).

□

5. Kinodynamic Motion Planning Framework

The motion planning structure comprises of four stages: (i) dynamic planning; (ii) Q-learning; (iii) terminal state evaluation; and (iv) obstacle augmentation, as shown in Fig. 1. The path planning is assigned to RRT$^\mathrm{X}$ which provides all the waypoints for navigation in a dynamic environment. Next, the online Q-learning and the terminal state evaluation are implemented. In parallel, we monitor the kinodynamic distance to augment the obstacle space. The implementation is presented in Algorithm 1.

Dynamic Planning: The RRT$^\mathrm{X}$ contains not only the sub-tree—which stores the desired path—but also the search-graph of the initial planning process. In this way, the algorithm reuses the search-graph for a rewiring cascade whenever the domain changes. Consequently, information is transferred rapidly throughout the tree in the modified environment. Moreover, RRT$^\mathrm{X}$ maintains an ϵ-consistent[1] graph, which guarantees the quality of existing paths

---

[1]ϵ-consistency means that the cost-to-goal value is within ϵ of the minimum of sum distance-to-neighbor and the neighbor's cost-to-go.

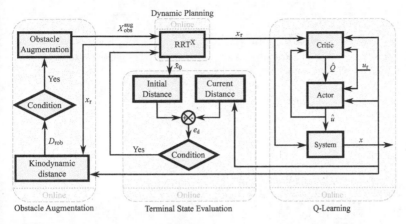

Figure 1: The motion planning structure consists of four stages. The algorithm runs clockwise, starting from the dynamic planning.

and allows for quick replanning. The neighborhood size at each node remains constant by selecting neighbors to maintain the runtime at each iteration. Therefore, RRT^X can provide a quick reaction in unpredictable dynamic environments as well as high-quality new paths.

Obstacle Augmentation: Since the model of the system is unknown, it is assumed that the robot traverses straight paths, given the waypoints from the RRT^X. In addition, optimality in terms of path planning usually indicates narrow distance between the obstacles and the path. In our case, kinodynamic constraints (1) as well as the optimal performance (2) result in traversing curved paths, instead of the assumed straight-line paths. Thus, there exists a deviation from the assumed straight-line path and the traversed path of the robot. This deviation of paths may result in unsafe navigation with collisions. To address this problem, we introduce the concept of kinodynamic distance and follow an obstacle augmentation strategy (Kontoudis and Vamvoudakis, 2019a,b). Therefore, instead of considering the physical shape of the obstacles, their augmented shape is taken into account. The augmented obstacle space $\mathcal{X}_{\text{obs}}^{\text{aug}}$ is computed trough the Minkowski sum (7) based on the maximum kinodynamic distance $D_{\text{rob}}^{\text{kin}}$. Whenever new obstacles are detected, the obstacle augmentation precedes the replanning process to avoid collision.

Q-Learning: At every k-pair of waypoints $(x_{0,k}, x_{r,k})$ of the planned path π, the proposed control law (16) is implemented to drive the

Algorithm 1 RRT-Q^X

Input: T - finite horizon; Δt - resolution; M, R - cost weight matrices; $P(T)$ - fixed Riccati matrix; ρ - admissible window; x_{goal} - goal state; x_{start} - start state; $\mathcal{X}_{\text{obs}}$ - obstacle space; $\mathcal{X}$ - state space
Output: $\hat{u}$ - control

1:   $\alpha_{\text{a}}, \alpha_{\text{c}} \leftarrow$ Stability(M, R) (25);
2:   $\mathcal{X}_{\text{obs}}^{\text{aug}} \leftarrow \mathcal{X}_{\text{obs}}$;
3:   $D_{\text{rob}}, D_{\text{rob}}^{\text{kin}} \leftarrow 0$; $k \leftarrow 1$;
4: **while** $x_{\text{goal}} \neq x$ **do**
5:     **while** *NoCollision* **do**
6:       $D_0 \leftarrow$ InitialDistance(x_0) (4);
7:       **for** $t \in T$ **do**
8:         **if** $D_{\text{rob}} > D_{\text{rob}}^{\text{kin}}$ **then**   ▷ Obstacle augmentation
9:           $D_{\text{rob}}^{\text{kin}} \leftarrow D_{\text{rob}}$;
10:          $\mathcal{X}_{\text{obs}}^{\text{aug}} \leftarrow$ Augment$(\mathcal{X}_{\text{obs}}, D_{\text{rob}}^{\text{kin}})$ (7);
11:         **end if**
12:         $\hat{W}_{\text{c}} \leftarrow$ Critic$(M, R, \Delta t, \alpha_{\text{c}}, \bar{x}, \hat{u})$ (22);   ▷ Q-learning
13:         $\hat{Q} \leftarrow$ EstimateQ$(\hat{W}_{\text{c}}, \bar{x}, \hat{u})$ (15);
14:         $\hat{W}_{\text{a}} \leftarrow$ Actor$(\hat{Q}, \alpha_{\text{a}}, \bar{x})$ (23);
15:         $\hat{u} \leftarrow$ Control$(\hat{W}_{\text{a}}, \bar{x})$ (16);
16:         **return** $\hat{u}$;
17:         $D_{\text{rob}} \leftarrow$ KinodynamicDistance$(x_{0,k}, \bar{x}, D_0)$ (6);
18:         **if** $D \leq \rho D_0$ **then**   ▷ Terminal state evaluation
19:           $x_{0,k} \leftarrow x(t)$;
20:           $k \leftarrow k + 1$;
21:           break;
22:         **end if**
23:       **end for**
24:     **end while**
25:     $\mathcal{G}, \pi \leftarrow$ RRT$^X(\mathcal{X}, \mathcal{X}_{\text{obs}}^{\text{aug}}, x_{\text{start}}, x_{\text{goal}})$;   ▷ Dynamic planning
26: **end while**

system. The critic is used to assess the policy, and the actor to perform the policy update. The critic approximates the Q-function according to (15), where $\hat{W}_{\text{c}}$ are the critic parameters that can be computed online by (22). The actor approximates the optimal control policy according to (16), where $\hat{W}_{\text{a}}$ are the actor parameters following the tuning law (23).

Terminal State Evaluation: A distance metric is employed to evaluate the terminal condition. At every Δt we compute the initial distance D_0 (4) and the relative distance D (5). When the relative distance drops below an admissible portion of the initial distance $D \leq \rho D_0$, then the algorithm assigns the current state as the new initial state $x_{0,k+1} = x(t)$ and proceeds to the next $(k+1)$-pair of waypoints.

Remark 2. The obstacle augmentation is a conservative strategy, because we are using the maximum kinodynamic distance to limit the free space. This means we are reserving space based on the worst case scenario that may appear only few times during the navigation. However, since we tackle the model-free problem without offline trials, that is a feasible methodology to ensure collision-free navigation.

6. Simulations and Results

In this section, we demonstrate the efficiency of the proposed motion planning technique in a dynamic environment with unpredictably appearing obstacles. We consider the continuous-time linear time-invariant system,

$$
\begin{bmatrix} \dot{x} \\ \dot{y} \\ \ddot{x} \\ \ddot{y} \end{bmatrix} = \begin{bmatrix} 0 & 0 & 1 & 0 \\ 0 & 0 & 0 & 1 \\ -.5 & 0 & -1.125 & 0 \\ 0 & -.5 & 0 & -1.125 \end{bmatrix} \begin{bmatrix} x \\ y \\ \dot{x} \\ \dot{y} \end{bmatrix} + \begin{bmatrix} 0 & 0 \\ 0 & 0 \\ .025 & 0 \\ 0 & .025 \end{bmatrix} \begin{bmatrix} f_1 \\ f_2 \end{bmatrix},
$$

(26)

where x, y is the translation, $\dot{x}, \dot{y}$ the velocities, and $\ddot{x}, \ddot{y}$ the acceleration along the x and y axes respectively. The inputs forces are denoted f_1, f_2. The system in (26) represents an autonomous rover. More details about the parameters of (26) are discussed in (Kontoudis and Vamvoudakis, 2019a).

The autonomous rover has full state feedback and limited perception range. Thus, an accurate configuration of the obstacles can be detected only in the perception range. The environment is completely unknown and consists of obstacles that appear throughout the navigation. Since the environment is unknown, we suppose there are no obstacles other than the obstacles detected in the perception range. The proposed method measures the kinodynamic distance D_{rob} and updates the augmented obstacle space $\mathcal{X}_{\mathrm{obs}}^{\mathrm{aug}}$ at every time instance Δt. We set the finite horizon as $T = 10\,\mathrm{s}$ for every run and the admissible window $\rho = 0.15$. The user-defined matrices are $M = 10 I_4$, and $R = 2 I_2$. The final Riccati matrix is $P(T) = 0.5 I_4$. We set the actor and critic gains as $\alpha_{\mathrm{c}} = 90$, and $\alpha_{\mathrm{a}} = 1.2$ respectively, by following (25). The resolution is $\Delta t = 0.05\,\mathrm{s}$. The initial values of the critic estimator vector $\hat{W}_{\mathrm{c}}$ and the actor esti-

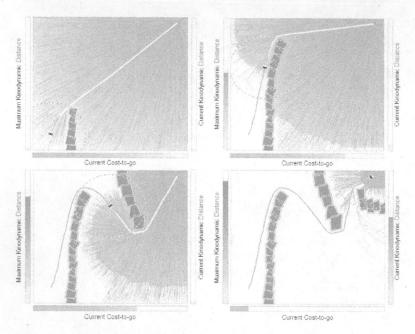

Figure 2: Various time frames of the autonomous rover collision-free navigation in an unpredictable dynamic environment.

mator vector $\hat{W}_a$ are randomly selected, except the last three elements of $\hat{W}_c$ that need to be non-zero. These elements are the $\{\hat{W}_c\}_{19:21} = Q_{uu}$ values that are inverted in (14). Note that there are three elements, because the user defined matrix R is symmetric and we are also employing the half-vectorization operation in (15). For the implementation of the RRT^X we use the package in (Otte, 2016).

The simulation is shown in Fig. 2, and a demonstrating video is available online[2]. The environment is a square with corners $(-20, -20)$ and $(20, 20)$ in meters. The start state is $x_{start} = (-15, -15, 0, 0)$ and the goal state $x_{goal} = (17, 17, 0, 0)$. The perception range is omnidirectional with a radius of $8\,\mathrm{m}$ and is illustrated with a dashed circle. The traversed path of the rover is shown in a solid line and the RRT^X path in a white line. The gray solid lines represent the search-tree of RRT^X. In every BVP, the rover moves toward the goal state. The initial shape of the obstacles is denoted by polygons and the corresponding augmentation in

---

[2]https://youtu.be/vNvOMTzxd0c.

light gray. The colored background represents the cost-to-go from every location to the final goal. Note that the cost-to-go is an underestimated value, as it is calculated with respect to straight-line paths. The maximum kinodynamic distance D_{rob}^{kin}, the current kinodynamic distance D_{rob}, and the cost-to-go of RRT^X are shown in the left, right and bottom bars, respectively. The autonomous rover successfully avoids the obstacles throughout the navigation in an unpredictable dynamic environment using the proposed kinodynamic motion planning technique.

7. Conclusion

This paper proposed a real-time kinodynamic motion planning methodology for unpredictable dynamic environments. More precisely, we introduced a Q-learning control law to approximate the optimal policy of a continuous-time linear time-invariant system and we used a terminal state evaluation and an obstacle augmentation technique. We rigorously derived the Q-learning controller, so that global asymptotic stability of the equilibrium point is ensured. The simulations reveal that the autonomous rover can efficiently perform safe navigation with no collisions in an unknown dynamic domain. Our methodology is completely model-free without offline training and requires insignificant computations that facilitate the execution of the algorithm in real-time.

Funding

This work was supported in part, by NASA ULI, and by NSF under grant Nos. CAREER CPS-1851588 and S&AS 1849198.

References

Allen, R. E. and Pavone, M. (2019). A real-time framework for kinodynamic planning in dynamic environments with application to quadrotor obstacle avoidance. Robotics and Autonomous Systems, 115: 174–193.

Berkenkamp, F. and Schoellig, A. P. (2015, July). Safe and robust learning control with Gaussian processes. pp. 2496–2501. In IEEE European Control Conference (ECC).

Bruce, J. and Veloso, M. M. (2002, June). Real-time randomized path planning for robot navigation. pp. 288–295. In Robot Soccer World Cup. Springer, Berlin, Heidelberg.

Bryson, A. E. (1975). Applied Optimal Control: Optimization, Estimation and Control. CRC Press.

Busoniu, L., Babuska, R., De Schutter, B. and Ernst, D. (2010). Reinforcement Learning and Dynamic Programming using Function Approximators (Vol. 39). CRC Press.

Chiang, H. T. L., Hsu, J., Fiser, M., Tapia, L., Faust, A. et al. (2019). RL-RRT: Kinodynamic motion planning via learning reachability estimators from RL policies. IEEE Robotics and Automation Letters, 4(4): 4298–4305.

Donald, B., Xavier, P., Canny, J. and Reif, J. (1993). Kinodynamic motion planning. Journal of the ACM, 40(5): 1048–1066.

Ferguson, D., Kalra, N. and Stentz, A. (2006, May). Replanning with RRTs. pp. 1243–1248. In IEEE International Conference on Robotics and Automation.

Hsu, D., Kindel, R., Latombe, J. C. and Rock, S. (2002). Randomized kinodynamic motion planning with moving obstacles. International Journal of Robotics Research, 21(3): 233–255.

Ioannou, P. A. and Sun, J. (2012). Robust Adaptive Control. Courier Corporation.

Karaman, S. and Frazzoli, E. (2011). Sampling-based algorithms for optimal motion planning. International Journal of Robotics Research, 30(7): 846–894.

Kavraki, L. E., Svestka, P., Latombe, J. C. and Overmars, M. H. (1996). Probabilistic roadmaps for path planning in high-dimensional configuration spaces. IEEE Transactions on Robotics and Automation, 12(4): 566–580.

Kontoudis, G. P. and Vamvoudakis, K. G. (2019). Kinodynamic motion planning with continuous-time Q-learning: An online, model-free, and safe navigation framework. IEEE Transactions on Neural Networks and Learning Systems, 30(12): 3803–3817.

Kontoudis, G. P. and Vamvoudakis, K. G. (2019, July). Robust kinodynamic motion planning using model-free game-theoretic learning. pp. 273–278. In American Control Conference.

Kontoudis, G. P., Xu, Z. and Vamvoudakis, K. G. (2020, July). Online, model-free motion planning in dynamic environments:

An intermittent, finite horizon approach with continuous-time Q-learning. pp. 3873–3878. In American Control Conference.

Kuffner, J. J. and LaValle, S. M. (2000, April). RRT-connect: An efficient approach to single-query path planning. pp. 995–1001. In IEEE International Conference on Robotics and Automation.

LaValle, S. M. (1998). Rapidly-exploring random trees: A new tool for path planning.

LaValle, S. M. and Kuffner Jr, J. J. (2001). Randomized kinodynamic planning. International Journal of Robotics Research, 20(5): 378–400.

Li, Y., Cui, R., Li, Z. and Xu, D. (2018). Neural network approximation based near-optimal motion planning with kinodynamic constraints using RRT. IEEE Transactions on Industrial Electronics, 65(11): 8718–8729.

Lewis, F. L., Vrabie, D. and Syrmos, V. L. (2012). Optimal Control. John Wiley & Sons.

Lewis, F. L., Vrabie, D. and Vamvoudakis, K. G. (2012). Reinforcement learning and feedback control: Using natural decision methods to design optimal adaptive controllers. IEEE Control Systems Magazine, 32(6): 76–105.

Lillicrap, T. P., Hunt, J. J., Pritzel, A., Heess, N., Erez, T. et al. (2015). Continuous control with deep reinforcement learning. arXiv preprint arXiv:1509.02971.

Mehta, P. and Meyn, S. (2009, December). Q-learning and Pontryagin's minimum principle. pp. 3598–3605. In IEEE Conference on Decision and Control.

Nägeli, T., Alonso-Mora, J., Domahidi, A., Rus, D., Hilliges, O. et al. (2017). Real-time motion planning for aerial videography with dynamic obstacle avoidance and viewpoint optimization. IEEE Robotics and Automation Letters, 2(3): 1696–1703.

Netter, J., Kontoudis, G. P. and Vamvoudakis, K. G. (2021, December) Bounded rational RRT-QX: Multi-agent motion planning in dynamic human-like environments using cognitive hierarchy and Q-learning. In IEEE Conference on Decision and Control.

Otte, M. and Frazzoli, E. (2014). RRT^X: Real-time motion planning/ replanning for environments with unpredictable obstacles. Algorithmic Foundations of Robotics, pp. 461–478.

Otte, M. and Frazzoli, E. (2016). RRT^X: Asymptotically optimal single-query sampling-based motion planning with quick replanning. International Journal of Robotics Research, 35(7): 797–822.

Otte, M. (2016). RRT-X (dynamic obstacles). Retrieved from: http:// ottelab.com/html_stuff/code.html#RRTXcode.

Perez, A., Platt, R., Konidaris, G., Kaelbling, L., Lozano-Perez, T. et al. (2012, May). LQR-RRT*: Optimal sampling-based motion planning with automatically derived extension heuristics. pp. 2537–2542. In IEEE International Conference on Robotics and Automation.

Powell, W. B. (2007). Approximate Dynamic Programming: Solving the Curses of Dimensionality (Vol. 703). John Wiley & Sons.

Sutton, R. S. and Barto, A. G. (2018). Reinforcement Learning: An Introduction. MIT Press.

Tedrake, R., Manchester, I. R., Tobenkin, M. and Roberts, J. W. (2010). LQR-trees: Feedback motion planning via sums-of-squares verification. International Journal of Robotics Research, 29(8): 1038–1052.

Vamvoudakis, K. G. (2017). Q-learning for continuous-time linear systems: A model-free infinite horizon optimal control approach. Systems & Control Letters, 100: 14–20.

Vrabie, D., Vamvoudakis, K. G. and Lewis, F. L. (2013). Optimal Adaptive Control and Differential Games by Reinforcement Learning Principles (Vol. 2). IET.

Watkins, C. J. and Dayan, P. (1992). Q-learning. Machine Learning, 8(3-4): 279–292.

Webb, D. J. and Van Den Berg, J. (2013, May). Kinodynamic RRT*: Asymptotically optimal motion planning for robots with linear dynamics. pp. 5054–5061. In IEEE International Conference on Robotics and Automation.

Yang, G. Z., Bellingham, J., Dupont, P. E., Fischer, P., Floridi, L. et al. (2018). The grand challenges of science robotics. Science Robotics, 3(14).

Chapter 2

An Ultrasound-Guided Mechatronics-Assisted System for Semi-Automated Seed Implantation and Tracking in Prostate Brachytherapy

Carlos Rossa,[1,*] *Jay Carriere,*[2] *Mohsen Khadem,*[3]
Ronald Sloboda,[4] *Nawaid Usmani*[4] and *Mahdi Tavakoli*[5]

1. Introduction

Prostate cancer is the most prevalent non-skin malignancy in men worldwide. Among the various treatment options that are available including surgery, external beam radiotherapy, and hormone therapy, transperineal interstitial permanent prostate brachytherapy (TIPPB) using radioactive seeds has emerged as an efficacious, minimally-invasive, patient-friendly, and cost-effective treatment option for localized prostate cancer. The robust clinical outcomes reported in the prostate brachytherapy literature highlight the great efficacy of this treatment modality when employed at cen-

[1] Department of Computer and Systems Engineering, Carleton University, Ottawa, ON, Canada.
[2] Department of Electrical and Computer Engineering, University of Calgary, Calgary, AB, Canada. Email: jay.carriere@ucalgary.ca
[3] School of Informatics, University of Edinburgh, Edinburgh, United Kingdom. Email: mohsen.khadem@ed.ac.uk
[4] Cross Cancer Institute and the Department of Oncology, University of Alberta, Canada. Emails: ron.sloboda; nawaid.usmani@albertahealthservices.ca
[5] Department of Electrical and Computer Engineering, University of Alberta, Edmonton, AB, Canada. Email: mahdi.tavakoli@ualberta.ca
* Corresponding author: rossa@sce.carleton.ca

tres performing high quality implants. However, brachytherapy is a technical procedure that relies on surgeons with sufficient expertise and case volume to maintain the quality of implants required to achieve these results.

Close scrutiny of TIPPB's technical aspects indicates room for considerable improvement. Contemporary brachytherapy techniques treat the whole prostate and involve the insertion of needles through the perineum using a template grid under ultrasound guidance. These seed-carrying needles are manually guided toward planned locations in the prostate, where the seeds are deposited, assuming that the needles will remain parallel across the entire length of their insertion. However, in practice, this assumption does not hold particularly well, causing the actual needle trajectories to not pass through the planned locations. Contributing to this error are prostate deformation/motion during needle insertion, imaging limitations, needle placement uncertainty Nath et al. (2000); Webster et al. (2006); Khadem et al. (2016), prostate swelling during implantation Sloboda et al. (2010), and seed migration Usmani et al. (2011). Experienced physicians can place seeds with an average absolute accuracy of no better than 5 mm, a substantial error of more than 10% of the average prostate diameter Taschereau et al. (2000).

Due to the currently limited accuracy of delivering seeds, brachytherapy has been limited to primarily treating the entire prostate gland for patients with localized prostate cancer. Treating the whole gland may result in side-effects such as urinary and rectal toxicity due to the effects on the adjacent structures. In the near future, anticipated focal treatment of dominant intraprostatic lesions identified by cancer-specific Magnetic Resonance Imaging (MRI) and Positron Emission Tomography (PET) imaging will require that seed placement accuracy must improve substantially. Improving source placement can result in enhanced treatment of localized prostate cancer by brachytherapy, and in addition will make this treatment modality applicable to other clinical situations. It has been estimated that between one-half to two-thirds of men with early stages of prostate cancer may be amenable to focal therapy Karavitakis et al. (2011); Bott et al. (2010).

To improve seed placement accuracy, robotics assisted needle steering and seed implantation have been proposed Podder et al. (2014); Muntener et al. (2006); Patriciu et al. (2007); Cowan et al. (2011); Rossa et al. (2016); Phee et al. (2006); Salcudean et al. (2008); Wei et al. (2004). To steer the seed-carrying needles, these systems either rotate the needle base on a measured basis during insertion, or apply forces at the base in order to control the needle tip's trajectory. The needle insertion can be performed manually Schneider et al. (2004) or automatically Muntener et al. (2006); Patriciu et al. (2007) while the robot orients the needle inside the tissue Schneider et al. (2004). In term of the degree of automation, these systems can essentially be classified into three main categories:

- *Fully automated steering*: The system automatically performs the needle insertion and the seed deposition in tissue Patriciu et al. (2007); Muntener et al. (2006); Phee et al. (2006); Meltsner et al. (2007); Adebar et al. (2011); Hungr et al. (2012); Yu et al. (2006); Phee et al. (2006). Although high accuracy can be achieved, integrating these systems with current clinical practice is challenging and most often, several modifications to the clinical setting are necessary.

- *Semi-automated steering*: The robotic system acts as a needle holder that either rotates the needle axially or manipulates the needle shaft Wei et al. (2004); Fichtinger et al. (2008); Schneider et al. (2004); Salcudean et al. (2008) with the physician being in charge of the insertion procedure. This category includes teleoperated needle insertion schemes Seifabadi et al. (2012).

- *Fully manual steering*: This class comprises technologies designed to provide the physician with relevant information about the necessary manoeuvres and keeps her/him in control of both insertion and steering procedures, such as visual and tactile feedback devices Rossa et al. (2016); Magee et al. (2007); Basu et al. (2016).

The first two categories often make use of complex structures that need to be integrated with the current clinical setting. In the third category, the implant outcomes still depend on the surgeon's ability to perform the necessary steering actions.

In Rossa et al. (2016), we introduced a new twist on robotic-assisted needle steering that uses a fully hand-held apparatus for accurate needle steering (see Fig. 1(a)). The device automatically rotates the needle at appropriate insertion depths as the surgeon manually inserts it. The system was designed to be entirely compatible with the current operating room setting, and hence, does not rely on any complex structures. In this paper, we extend this new framework to accurate seed implantation and tracking using ultrasound images in focal low dose rate brachytherapy.

Contributions of this paper include: (i) The device is modified to incorporate a miniature force sensor that measures the tissue parameters required in the needle steering controller online. (ii) A new needle steering controller based on the Rapidly Exploring Random Tree algorithm is implemented, and (iii) a method is devised to track the position of each implanted seed on-line. The concept is validated by implanting dummy seeds in biological and synthetic tissue samples in order to achieve a hypothetical desired seed distribution. Experimental results obtained from 90 seed implants indicate an accuracy of 0.46 mm in delivering the seeds. This is the first implementation of a fully hand-held seed implantation and tracking system for the emerging modality of focal prostate cancer treatment.

2. The Hand-Held Apparatus for Seed Implant

In order to perform needle insertion and seed deposition, we modified the needle steering device previously presented in Rossa et al. (2016) (for a video please see https://goo.gl/Z7jJp5) (see Fig. 1(a)). Standard brachytherapy needles are connected to the apparatus, which can rotate the needle base axially (see Fig. 1(b)). As the surgeon uses the device to insert the needle, the 3D position of the apparatus is measured in real time by an optical motion tracker that follows markers placed on the side of the device (not visible in Fig. 1(a)). An important difference compared to the device presented in Rossa et al. (2016) is that in this paper, a compression/traction sensor (model LSB200 S-Beam from Futek, Irvine, USA) is embedded in the device in order to measure the axial force applied to the needle base during insertion and withdrawal. The goal is to employ the force measurements from the two 1-DOF force sensors during needle insertion and withdrawal to estimate the forces applied by the tissue onto the needle tip, such that future needle deflection can be predicted by a mechanics-based model and the necessary corrective action taken by the hand-held apparatus.

To simulate radioactive brachytherapy seeds, the dummy seeds made of 5 mm long, 1 mm in diameter, stainless steel cylinders, shown in Fig. 1(c), were fabricated. A single seed and a stylet are loaded in the needle. Once the needle reaches the desired depth, the

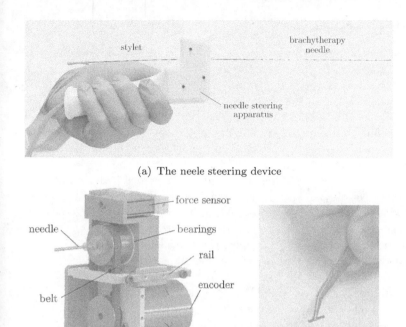

(a) The neele steering device

(b) Actuation unit (c) Dummy seeds

Figure 1: In (a), the needle steering device introduced in Rossa et al. (2016) is shown. (b) presents the upgraded actuation unit that comprises a 1-DOF force sensor that measures the needle insertion and withdrawal forces. The device steers the needle and is used to deposit the dummy seeds shown in (c).

surgeon holds the stylet in place and withdraws the needle (with the device) such that the stylet removes the seed from the needle shaft for deposition in tissue.

3. Needle Steering Controller

This section presents the model and steering algorithm that are combined to steer the needle towards the target.

3.1 Needle-Tissue Modelling

We will employ the needle-tissue interaction model we presented in Rossa et al. (2016). The inputs to the model are the needle insertion depth, the current needle deflection, and the rotation depth(s). The model outputs the future needle tip trajectory and the needle shape. In Rossa et al. (2016), the needle is modelled as a cantilever compliant beam that experiences forces applied by the tissue in the form of 1) a needle-tissue cutting force, which is applied at the needle tip normal to the needle shaft (denoted by F), and 2) the tissue reaction force applied along the needle shaft, which is due to compression of the tissue. This latter force depends on the stiffness of tissue per unit length of the needle (denoted by K).

In Rossa et al. (2016), a method is devised to estimate both F and K based only on ultrasound images. However, this requires the ultrasound probe to follow the needle during insertion. Thanks to the force sensor added to the actuation unit, F can be measured without need for image feedback. Once F is determined, K can be found by fitting the model to a single deflection measurement point obtained from a stationary ultrasound probe.

In order to calculate the force F applied at the needle tip, the needle steering apparatus measures the forces applied to the needle's base F_{in} that are necessary to insert and withdraw it from the tissue (see 2(a)). From this information, we will derive F by following the procedure shown in Fig. 2. In Fig. 2(a), as the needle is pushed into tissue, a force F_c is applied at the needle tip, that has transverse and longitudinal components Q, and F, respectively. These forces are functions of F_c and of the needle bevel angle β. As the surgeon pushes the needle into the tissue, the measured force at the needle base F_{in} corresponds to $F_1 = P + f$ where f is the needle-tissue frictional force along the shaft given by $f = (bv_1)d$, where v_1 is the insertion velocity, and b is the friction coefficient per unit length of the inserted needle. When the needle is withdrawn after insertion, the measured force F_2 corresponds to friction only. If the needle is withdrawn with a velocity of v_2, the force P can be found as

$$P = F_1 - F_2 \left(\frac{v_1}{v_2} \right) \qquad (1)$$

It is thereby implied that b is constant during insertion and withdrawal Khadem et al. (2016). The force F is finally computed as $F = P(\tan \beta)^{-1}$, where β is the needle bevel angle. Knowing F, one can determine K by fitting the model such that the estimated

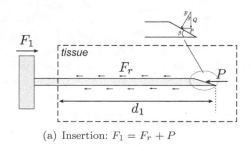

(a) Insertion: $F_1 = F_r + P$

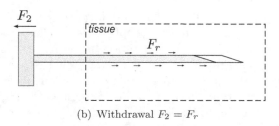

(b) Withdrawal $F_2 = F_r$

Figure 2: Needle insertion in soft tissue. As the needle cuts a path in the tissue, a force F_c is applied to the needle tip. The horizontal component of F_c plus friction along the shaft correspond to the needle insertion force measured at the needle base. During needle withdrawal, the measured force corresponds to friction only.

needle deflection $\hat{v}_i(K)$ matches the measured deflection v_i of an inserted needle, at a point i along its shaft. More specifically, K is found to minimize

$$J(K) = \min \sum_{i=1}^{n} (v_i - \hat{v}_i(K))^2, \tag{2}$$

where n is the number of measurements taken.

Once the needle-tissue model parameters are identified, the model can be used to estimate the optimal needle rotation depths as described in the next subsection.

3.2 Needle Steering Controller

In this section, a novel motion planner is developed. The motion planner computes a large number of needle tip trajectories (plans) using the model presented in Rossa et al. (2016) and selects the best plan. It outputs a set of depths at which the needle is axially rotated that brings the needle to the target. The planner uses the Rapidly Exploring Random Tree (RRT) algorithm LaValle and Kuffner (2001); Patil et al. (2014) to calculate the rotation depths. RRT is an efficient sampling algorithm to quickly search high-dimensional spaces that have algebraic constraints such as the number of allowed needle rotations, by randomly building a space-filling tree. Figure 3(a) shows a block diagram of the closed-loop control algorithm based on the online motion planning.

To design the online motion planner we present the needle steering problem in the needle configuration space, called $\mathcal{C}$. Assuming the needle moves in a 2D insertion plane, the needle workspace is a Euclidean space $\mathcal{W} = \mathbb{R}^2$. The configuration space ($\mathcal{C}$) is the space of all possible control actions (i.e., depth(s) of needle rotation(s)), whose values identify the configuration of the needle tip in the workspace. Considering symmetry of rotation depths (e.g., rotations at depths of 40 and 80 mm are equal to rotations at 80 and 40 mm) the configuration space is an n-dimensional simplex, where n is the number of axial rotations. For instance, if the maximum allowable number of rotations is 3, the configuration space forms a tetrahedron.

The proposed motion planner uses an approximate decomposition of $\mathcal{C}$. Assuming that the distance between two consecutive rotations is at least 5 mm, $\mathcal{C}$ can be decomposed into several smaller simplices shown in Fig. 3(c). This is a valid assumption since two close 180° axial rotations are equal to one 360° rotation of the needle tip and this action has no effect on needle deflection.

The disjoint cells in $\mathcal{C}$ form a *connectivity graph*. The nodes of this graph are vertices of the cells corresponding to a certain configuration (i.e., rotation depths). Assuming that the initial guess for a configuration in $\mathcal{C}$ is q_s and the goal configuration that steers the needle toward the target is q_g, planning a motion for the needle involves searching the connectivity graph for a path from cell containing q_s to the cell containing q_g. For this purpose we use the RRT algorithm. In the following a pseudocode description of the motion planner algorithm is given.

The inputs of the RRT are the current depth X_0, the number of allowed rotations N, and the computation time available for planning T_{max}. A hypothetical example of tree generation for $N = 2$ is shown in Fig. 3(b). First, the configuration space $\mathcal{C}$ is formed based on the number of allowed rotations N and the current needle insertion depth $X_0 = 0$. The tree is initialized with a first vertex q_s located at $(0, 0)$ (see (I) in Fig. 3(b)). The algorithm then generates a random candidate q_{rand} from the N-dimentional

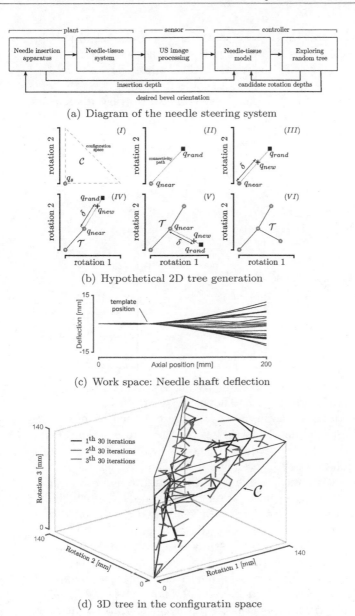

(a) Diagram of the needle steering system

(b) Hypothetical 2D tree generation

(c) Work space: Needle shaft deflection

(d) 3D tree in the configuratin space

Figure 3: Overview of needle steering controller. (a) shows the block diagram of the needle steering system. In (b), the RRT algorithm evaluates the needle targeting accuracy for different rotation depths as shown in (c). In (d) the resultant set of rotation depths.

configuration space C (See Rand_Conf in Algorithm 1 and (II) in Fig. 3(b)). Next, Near_Vertex runs through all the vertices (candidate rotation depths) in C to find the closest vertex to q_{rand}. New_Conf produces a new candidate configuration q_{new} on the segment joining q_{near} to q_{rand} at a predefined arbitrary distance δ from q_{near} (see (III) in Fig. 3(b)).

Algorithm 1: q_{goal} $\leftarrow$ RRT_Algorithm (X_0, N, T_{max})

C $\leftarrow$ Initialize_space (X_0, N)
$\mathcal{T}$ $\leftarrow$ Initialize_tree $(X_\bullet, N)$
while $\chi = \varnothing \wedge \Gamma < \Gamma_{max}$ **do**
  q_{rand} $\leftarrow$ Rand_Conf (C)
  q_{near} $\leftarrow$ Near_Vertex (q_{rand}, C)
  q_{new} $\leftarrow$ New_Conf (q_{rand}, q_{near})
  p_{new} $\leftarrow$ Needle-tissue-model *Rossa et al. (2016)*
  (q_{new})
  $\mathcal{T}$ $\leftarrow$ Add_Vertex (q_{new})
  $\mathcal{T}$ $\leftarrow$ Add_Edge (q_{new}, q_{near})
  if $p_{new} \in \mathcal{G}$ **then**
    | q_{goal} $\leftarrow$ Extract_Conf (q_{new})
  end
end

The random tree $\mathcal{T}$ is expanded by incorporating q_{new} and the segment joining it to q_{near}, as shown in (VI). Next, the needle tip path and targeting accuracy (p_{new}) are obtained by inputting the selected rotation depths in the needle-tissue interaction model Rossa et al. (2016). The predicted needle shape for various candidate sets of rotation depths is shown in Fig. 3(c). When the needle path for the newly added configuration is found to lie in the target region ($\mathcal{G}$), or when the computation times exceeds T_{max} the RRT planner terminates. The target region is a closed circle with 1 mm diameter, centred on the desired target location in $\mathcal{W}$. The former condition implies that when the estimated needle tip deflection at the maximum depth is less than 0.5 mm, the algorithm stops. If the stopping condition is not met, the algorithm continues to expand the tree with new vertices as depicted in (V) and (VI) in Fig. 3(b).

Once the algorithm stops, the output q_{goal} contains the best set of rotation depths that will bring the needle towards $\mathcal{G}$. The RRT expansion procedure results in a very efficient exploration of C and the procedure for generating new candidates in RRT is intrinsically biased toward regions of C that have not been visited.

In prostate brachytherapy, the needle insertion point and the target are typically on the same horizontal line. Throughout this paper, we assume that the target is at a depth of 140 mm. In order to limit tissue trauma, the total number of needle axial rotations is set to three. Results of the simulation of the motion planner in configuration space $\mathcal{C}$ and the corresponding needle deflection predictions in needle workspace $\mathcal{W}$ for an insertion depth of 140 mm starting at 0 mm are shown in Fig. 3(c) and Fig. 3(d), respectively.

The RRT has been used for needle steering in LaValle and Kuffner (2001). Unlike LaValle and Kuffner (2001), our search space is directly constrained by the possible control inputs and by the number and depths of rotations. Therefore, there is no need to solve for the inverse kinematics of the model, which enables the optimization problem to be solved faster and makes the solution method suitable for online applications.

4. Experimental Setup

The setup for semi-automated seed implant is presented in Fig. 4. Please see the attached video. Standard 18-gauge clinical brachytherapy needles (Eckert & Ziegler Inc., USA) are loaded with a single dummy seed shown in Fig. 8(a) and connected to the needle steering apparatus. As in manual brachytherapy, a stylet is inserted in the needle shaft in order to deposit the seed in the tissue.

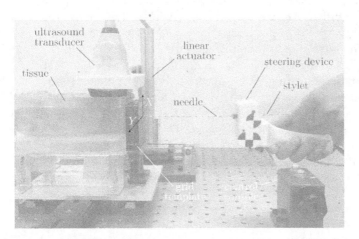

Figure 4: Experimental setup. A standard 18-gauge brachytherapy needle carrying a single dummy seed is inserted in the tissue through a guiding template. An ultrasound probe monitors the position of the needle tip.

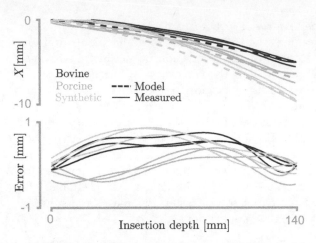

Figure 5: Model fit results for each tissue sample. The model parameters are found by minimizing the difference between the measured and estimated needle tip deflection at the depth of 140 mm.

The needle is inserted through a standard brachytherapy template grid (D0240018BK, from CR Bard, USA). For further details on the hardware implementation, we refer the reader to Rossa et al. (2016).

A 4DL14-5/38 linear ultrasound probe is placed on the tissue surface to acquire transverse 2D ultrasound images of the needle at 30 Hz. A linear stage motorized by a DC motor controls the position of the ultrasound probe using a discrete PID controller.

Three different tissues are used in the experiments. The first tissue is made by encasing a 130 mm long piece of porcine tissue into a mixture of 20% gelatin derived from acid-cured tissue (gel strength 300 from Sigma-Aldrich Corporation, USA) per litre of water. This tissue can be seen in Fig. 4. The gelatin is meant to create a 20 mm layer of tissue through which the needle is inserted before reaching the porcine tissue, and also to create a flat surface in order to ensure good acoustic contact between the ultrasound probe and the tissue. In the second tissue, the porcine layer is replaced with bovine tissue. Hence, the first two tissues are composed of two different layers. The third tissue is made of high friction plastisol gel (M-F Manufacturing Co., USA) mixed with 20% plastic softener.

For each tissue, 15 needle insertions at different locations in the grid template followed by deposition of a single seed are performed. The seeds are deposited at a depth of 140 mm. For each tissue, a set of 15 insertions is performed using an open loop controller (image feedback is not used), and another set of 15 implants is performed using a closed-loop needle insertion controller. This amounts to a total of 6 different experimental scenarios and 90 seed implants in total.

Each seed implantation procedure is composed of three phases:

1. *Phase 1–Pre-scan*: The needle has not been inserted in the tissue. The ultrasound moves with a constant velocity of 8 mm·s^{-1} up to a depth of 150 mm and returns to the initial position. Thereby, all previously implanted seeds and tracks in the tissue left by other insertions can be identified.

2. *Phase 2–Needle insertion*: The ultrasound imaging plane is placed close to the needle tip. During insertion, the ultrasound probe moves in synchrony such that the needle tip is

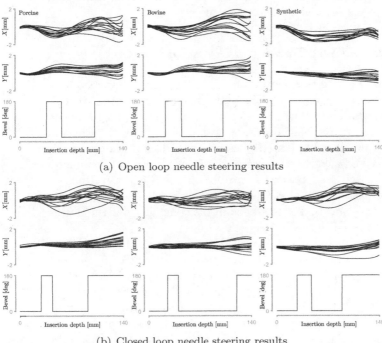

(a) Open loop needle steering results

(b) Closed loop needle steering results

Figure 6: Path followed by the needle tip in the X and Y planes (defined in Fig. 4) during insertion in porcine, bovine, and synthetic tissue and the average position of the bevel angle using open loop (a) and closed loop (b) controllers, for each of the 15 insertions. Only the deflection in the X plane is controlled.

always visible in the image. Once the needle reaches the desired depth of 140 mm, the seed is manually deposited and the needle is withdrawn.

3. *Phase 3–Post-scan*: After the needle is withdrawn the tissue is scanned in order to identify the position of the seed deposited in Phase 2.

The needle steering controller will be employed in two different ways. In *open-loop* mode, the controller determines 3 optimal rotation depths prior to needle insertion. In *closed-loop* mode, the RRT controller updates the rotation online based on the measured needle tip position. The maximum computation time allowed for planning is 1 second, which was found to provide good convergence. The needle bevel angle is initially oriented such that the needle deflects in a plane that is parallel to the table shown in Fig. 4. Deflection along the vertical plane is not controlled.

4.1 Needle and Seed Tracking in Ultrasound Images

Needle tip tracking is done online as the needle is inserted into the tissue. Each transverse ultrasound image is processed in real-time using the algorithm presented in Waine et al. (2016). Seed localization is done using the information from both the Phase 3 scan, containing the implanted seed, and the Phase 1 scan, which is used to reduce background noise in the Phase 3 transverse images. Final implanted seed positions are obtained offline after Phase 3 scan is completed. Note that when *open-loop* needle steering is used, the images are not used as feedback in the controller but the needle tip is still tracked.

From the final needle tip position in Phase 2, the seed deposition depth is obtained and the transverse ultrasound image that contains the seed can be selected from the Phase 3 scan, which we will denote as I_{P3}. The original image obtained in Phase 3 is shown in Fig. 7(b). Even with the deposition depth of the seed known, seed localization in transverse images is complicated by several factors, the most important of which is that previous seeds are present alongside the target seed, as well as the seed not being very distinct from the background image noise. An additional complication is that the implanted seed moves away from the final needle tip location, found in Phase 2, as the needle is withdrawn.

The seed tracking algorithm consists of 2 stages, i.e., a pre-processing stage and the background noise removal, see Fig. 7(a). The first step in the pre-processing stage is to define a region of interest (ROI) around the final needle tip location, found in Phase 2, in I_{P3} that is large enough to capture the seed with moderate motion. Empirically, an ROI of 100 px by 100 px is found to be sufficient. The next step is to find the ultrasound image at the seed deposition depth captured in Phase 1, which we will call I_{P1}. This image contains the previously deposited seeds as well as background noise from the phantom tissue. In order to remove the noise and other seeds from the ROI in I_{P3} the exact same ROI is taken from I_{P1} and the background is removed through a subtraction, such that a cleaner image, denoted I_C, is created, where $I_C = |I_{P3} - I_{P1}|$. The image I_C is then enhanced through the same contrast stretching method given in Waine et al. (2016), see Fig. 7(b).

With the background noise and previous seeds removed from the image, the target seed is now quite distinct from the background and so the final step is the seed segmentation. A straightforward binary threshold, determined empirically to count any pixel with an intensity above 150 (on a scale from 0 to 255). As a final segmentation step all 4-connected component objects in the binary image are found and the object with the largest number of pixels is chosen as the seed. The seed location is then determined by taking the x and y centroids of all of the pixels in the seed's 4-connected object. Please see the attached video or visit `https://youtu.be/tnWdMXSxmiU`.

5. Experimental Results

This section is divided into three main parts. First, calibration of the needle steering controller is presented. Next, the needle steering and seed implant results are shown.

5.1 Model Identification

The first step in performing assisted needle steering for accurate seed deposition is to calibrate the needle steering controller. To this end, 3 needle insertions followed by withdrawals are performed in each tissue at an average velocity of 2 mm·s[1]. The controller is turned off and the needle insertion/withdrawal force is recorded. For verification purposes, the ultrasound probe is following the needle tip. However, in a clinical scenario the ultrasound probe could

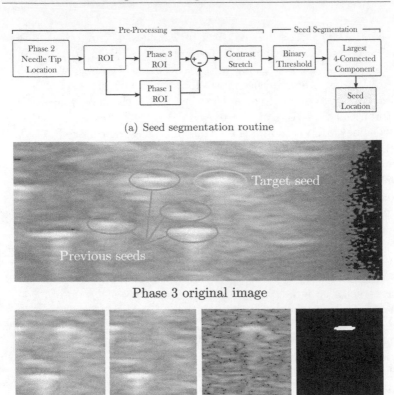

(a) Seed segmentation routine

Phase 3 original image

Phase 3 ROI   Phase 1 ROI   Enhanced   Binary

(b) Example of ultrasound image processing

Figure 7: Seed tracking routine in ultrasound images (a). The image processing is presented in (b). Ultrasound images captured during a Phase 3 showing the last implanted seed to be localized, with the tracking algorithm steps shown underneath.

instead be maintained stationary at the maximal insertion depth to measure the needle deflection at a single depth.

Following the procedure described in Section III, the force applied at the needle tip is identified. The obtained force is input to the needle-tissue interaction model Rossa et al. (2016) and the needle deflection is estimated for various candidate tissue stiffness values. The optimal needle-tissue stiffness is the one that minimizes the difference between the predicted and observed needle tip deflection at the maximal insertion depth. Figure 5 presents the results obtained with the identified model parameters. The prediction er-

Table 1: Indentified needle tip force (N), tissue stiffness (N·mm^{-2}), and average absolute prediction error (mm).

	Porcine tissue	Bovine tissue	Synthetic tissue
Force	1.10 ±0.07	1.26 ±0.05	0.78 ±0.12
Stiffness	72.6	86.5	36.6
Mean error	0.53 ±0.28	0.83 ±0.44	0.89 ±0.62

ror is less than 1 mm for all tissue samples. The results, including the optimal tissue stiffness, are summarized in Table 1.

5.2 Seed Implant with Non-image Based Needle Steering

Knowing all the parameters necessary for estimating the needle tip trajectory, the depths of rotation are determined by the controller. Let us first assume that no image feedback is available. Therefore, the controller is only used prior to the needle insertion. The needle is inserted through the grid template at different locations spaced 5 mm apart as in current clinical brachytherapy. 15 insertions are performed followed by seed deposition. The path followed by the needle tip is shown in Fig. 6(a) along with the orientation of the needle bevel angle. Over 45 insertions, the average needle targeting accuracy in the X and Y directions is 0.93 and 0.62 mm with the highest error occurring in bovine tissue and the lowest error observed in porcine tissue.

Once the needle reaches the depth of 140 mm, the seed loaded in the needle shaft is deposited in tissue and the needle is withdrawn. The final seed location with respect to the desired hypothetical seed distribution is shown in Fig. 8(a). The gray solid dot indicates the desired seed location, which is defined as a point in a 2D plane parallel to the grid template at a depth of 140 mm. The final needle tip location is shown by the blue circle and the square is centroid of each seed after needle withdrawal. The average seed targeting accuracy in the X and Y planes is 0.89 and 0.60 mm, respectively. During needle withdrawal the tissue deforms and moves the seeds by up to 0.30 mm see (Fig. 9). These results are summarized in Table 2.

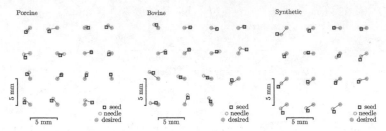

(a) Targeting accuracy and final seed location with open loop needle steering.

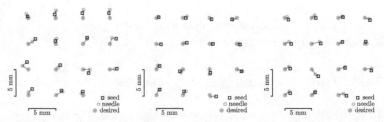

(b) Targeting accuracy and final seed location with closed loop needle steering.

Figure 8: Experimental results of seed deposition following a hypothetical pre-planning. The solid gray dot indicates the seed target location. The blue circle is the position of the needle tip at the target depth, and the dark square shows the final position of the centroid of each seed after the needle is withdrawn.

5.3 Seed Implant with Image-based Needle Steering

Let us now assume that the position of the needle tip can be measured at any time during insertion from ultrasound images. As a result, the steering controller can update the optimal rotation depths on-line. This is expected to result in an immediate improvement of targeting accuracy since the controller replans the path towards the target given the current position of the needle tip X_0, and the number n of axial rotations that have been performed.

The path followed by the needle tip is presented in Fig. 6(b). The third panel shows the *average* position of the bevel angle. The absolute needle targeting accuracy in the X and Y planes is 0.57 and 0.53 mm, respectively. Considering the deflection along X, this corresponds to an improvement of 40% compared to the case without image feedback. The final needle tip location at the target depth and the final location of the deposited seeds are shown in Fig. 8(b).

Table 2: Experiential results. Average absolute needle targeting accuracy, seed placement error and seed deviation after needle withdrawal, and average depth of needle rotation. Units are in millimetres.

		Porcine tissue	Bovine tissue	Synthetic tissue	Average
non-image based	X needle	0.69 ±0.45	1.07 ±0.41	1.05 ± 0.28	0.93
	Y needle	0.63 ±0.38	0.68 ±0.48	0.56 ± 0.38	0.62
	X seed	0.81 ±0.36	0.86 ±0.38	1.01 ± 0.46	0.89
	Y seed	0.53 ±0.30	0.46 ±0.37	0.81 ± 0.53	0.60
	X motion	0.27 ±0.34	0.40 ±0.37	0.29 ± 0.21	0.32
	Y motion	0.35 ±0.17	0.22 ±0.23	0.36 ± 0.23	0.31
	Rotation 1	31.1	18.7	12.8	
	Rotation 2	51.3	40.5	49.1	
	Rotation 3	100.9	102.5	118.9	
image based	X needle	0.51 ±0.44	0.39 ±0.26	0.81 ± 0.30	**0.57**
	Y needle	0.79 ±0.52	0.41 ±0.34	0.40 ± 0.25	0.53
	X seed	0.60 ±0.48	0.59 ±0.25	0.21 ± 0.89	**0.46**
	Y seed	0.84 ±0.34	0.34 ±0.29	0.31 ± 0.31	0.49
	X motion	0.38 ±0.24	0.31 ±0.26	0.21 ± 0.21	0.30
	Y motion	0.47 ±0.22	0.11 ±0.09	0.31 ± 0.31	0.29
	Rotation 1	39.2 ±12.4	36.8 ±9.3	38.2 ±7.7	
	Rotation 2	52.4 ±13.7	49.6 ±11.9	55.2 ±10.3	
	Rotation 3	98.5 ±16.4	122 ±15.8	95.8 ±12.2	

The average deviation from the actual to the desired seed location is 0.46 and 0.49 mm in the vertical and horizontal planes, respectively. The second part of Table 2 summarizes these results.

5.4 Discussion

Two different approaches have been proposed to steer a seed-carrying needle towards a pre-defined target. In the first approach the needle steering apparatus rotates the needle base at optimally depths determined preoperatively. In the second case, the the current position of the needle tip is used to update the optimal rotation depths intraoperatively.

The first method is compatible with a clinical setting where real-time measurement of the needle tip cannot be obtained during insertion. To address this limitation the steering apparatus is

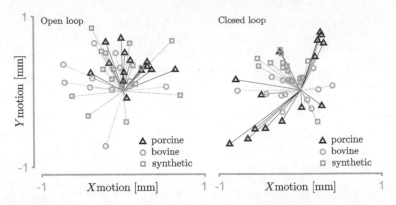

Figure 9: Dummy seed displacement from the deposition location during needle withdrawal in each tissue with open loop (left) and closed loop (right) needle steering controllers.

equipped with a force sensor that measures the needle insertion and withdrawal forces and estimates the required model parameters using the deflection measured at a single depth after insertion. 15 seeds are implanted 5 mm apart in the tissue to form a hypothetical seed distribution. The average needle and seed targeting accuracy in the controlled deflection direction is 0.93 and 0.89 mm on average, respectively.

The second method uses ultrasound images to measure the needle tip deflection in tissue as it is inserted. The controller running at 1 Hz recalculates the steering manoeuvres online, such that deviations from the offline predicted path can be corrected. With this approach, the average seed placement error is reduced to 0.46 mm.

Some commercially available ultrasound systems can be employed to follow the needle tip during insertion. Examples include the TargetScan from Envisioneering Medical, Overland, USA, where the 2D axial imaging plane translates within a stationary transrectal probe, and the 3D-2052 ultrasound probe from B&K Ultrasound. Peabody, USA, where the imaging plane translates axially by 70 mm. As an alternative, the Sonalis Ultrasound System from Best Medical, Pittsburgh, USA, has a longitudinal array that provides for 140 mm length of view, encompassing the bladder, the prostate and the perineum. Hence, the needle can be observed during throughout the insertion as long as it does not deflect out of the imaging plane.

Standards for seed implant quality are typically defined in terms of quantitative X-ray Computed Tomography-based postoperative dosimetric evaluation. Currently, ultrasound-based postoperative seed identification cannot be done routinely with any better than 80% accuracy Han et al. (2003); Wei et al. (2006). CT-based dosimetry evaluation requires a separate imaging session to scan the patient prostate in order to determine the final location of the seeds. This assessment is subject to anatomical variations of the prostate position and postoperative edema of the prostate gland. With the proposed method in this paper, assessment and corrections regarding seed implantation errors can be taken during the procedure without the need for postoperative imaging.

6. Conclusion

In this paper we demonstrate the feasibility of a new framework for accurate radioactive seed implantation and tracking during low dose rate prostate brachytherapy for prostate cancer. A hand-held needle steering apparatus controls the deflection of a seed-carrying needle during insertion such that the needle tip reaches the desired target with minimum deflection. The steering controller evaluates the effects of axial needle rotations at different depths on the needle targeting accuracy via a needle-tissue interaction model. Optimal rotation depths are determined prior to the procedure and can be updated as the needle insertion progresses. The device *automatically* steers the needle as the surgeon *manually* inserts it in tissue, keeping the surgeon in control of the procedure. Once the needle reaches the target, the surgeon can deposit the seeds in tissue as in current clinical practice. Hence, the proposed framework does not require major modifications to the operating room setup. Knowing the final needle tip location prior to seed deposition, a method is proposed to track the final seed locations after needle withdrawal, allowing the surgeon to monitor implant quality on the fly.

Despite the current clinical individual seed placement uncertainty of 5 mm, very good clinical results for brachytherapy can be achieved when the whole prostate gland is treated. This is a consequence of the large number of seeds involved in a whole gland implant (typically 80 to 100), and the addition of a 3 mm margin around the prostate to create a planning target volume to which the treatment dose is prescribed Salembier et al. (2007). With the proposed system, the average seed placement accuracy is improved to 0.46 mm in tissue phantoms. Reducing seed placement error to this order in the clinic can enable accurate brachytherapy boost or focal treatment of dominant intra-prostatic lesions rather than treating the whole prostate gland. Seeds carrying higher radiation doses can be considered to reduce the number of implanted seeds and the targeted areas within the prostate. Combined with improved imaging techniques Atri et al. (2009), it is possible to identify men with low- to intermediate-risk prostate cancer who have low volume focal disease and who may be suitable for local therapy. This would result in fewer side effects to the patient including reduced urinary problems, rectal symptoms, and improved erectile function Langley et al. (2012). In addition, the possibility of post-treatment after focal brachytherapy is expected to be easier than after conventional treatment of the whole prostate gland. Among the options for such treatment, it is possible to treat remaining regions of the prostate volume with specific techniques of external irradiation or salvage surgery Cosset et al. (2013).

Competing Interests

None to declare.

Funding

This work was supported by the Natural Sciences and Engineering Research Council (NSERC) of Canada (CHRP #446520), the Canadian Institutes of Health Research (CIHR) (CPG #127768), and by the Alberta Innovates - Health Solutions (AIHS) (CRIO #201201232).

References

Adebar, T., Salcudean, S., Mahdavi, S., Moradi, M., Nguan, C. et al. (2011). A robotic system for intra-operative transrectal ultrasound and ultrasound elastography in radical prostatectomy. In Information Processing in Computer-Assisted Interventions, pp. 79–89.

Atri, M., Gertner, M., Haider, M., Weersink, R., Trachtenberg, J. (2009). Contrast-enhanced ultrasonography for real-time monitoring of interstitial laser thermal therapy in the focal treatment of prostate cancer. Canadian Urological Association Journal, 3(2): 125–30.

Basu, S., Tsai, J. and Majewicz, A. (2016). Evaluation of tactile guidance cue mappings for emergency percutaneous needle insertion. In IEEE Haptics Symposium.

Bott, S., Ahmed, H., Hindley, R., Abdul-Rahman, A., Freeman, A. et al. (2010). The index lesion and focal therapy: An analysis of the pathological characteristics of prostate cancer. BJU International, 106(11): 1607–1611.

Cosset, J. -M., Cathelineau, X., Wakil, G., Pierrat, N., Quenzer, O. et al. (2013). Focal brachytherapy for selected low-risk prostate cancers: A pilot study. Brachytherapy, 12(4): 331–337.

Cowan, N., Goldberg, K., Chirikjian, G., Fichtinger, G., Alterovitz, R. et al. (2011). Robotic needle steering: Design, modeling, planning, and image guidance. In Surgical Robotics, pp. 557–582.

Fichtinger, G., Fiene, J., Kennedy, C., Kronreif, G., Iordachita, I. et al. (2008). Robotic assistance for ultrasound-guided prostate brachytherapy. Medical Image Analysis, 12(5): 535–545.

Han, B., Wallner, K., Merrick, G., Butler, W., Sutlief, S. et al. (2003). Prostate brachytherapy seed identification on post-implant trus images. Medical Physics, 30(5): 898–900.

Hungr, N., Baumann, M., Long, J. -A. and Troccaz, J. (2012). A 3D ultrasound robotic prostate brachytherapy system with prostate motion tracking. Robotics, IEEE Transactions on 28(6): 1382–1397.

Karavitakis, M., Winkler, M., Abel, P., Livni, N., Beckley, I. et al. (2011). Histological characteristics of the index lesion in whole-mount radical prostatectomy specimens: Implications for focal therapy. Prostate Cancer and Prostatic Diseases, 14(1): 46–52.

Khadem, M., Rossa, C., Sloboda, R. S., Usmani, N., Tavakoli, M. et al. (2016). Mechanics of tissue cutting during needle insertion in biological tissue. IEEE Robotics and Automation Letters, 1(2): 800–807.

Khadem, M., Rossa, C., Usmani, N., Sloboda, R. S., Tavakoli, M. et al. (2016). A two-body rigid/flexible model of needle steering dynamics in soft tissue. IEEE/ASME Transactions on Mechatronics, 99: 1–1.

Langley, S., Ahmed, H. U., Al-Qaisieh, B., Bostwick, D., Dickinson, L. et al. (2012). Report of a consensus meeting on focal low dose rate brachytherapy for prostate cancer. BJU International, 109(s1): 7–16.

LaValle, S. M. and Kuffner, J. J. (2001). Randomized kinodynamic planning. The International Journal of Robotics Research, 20(5): 378–400.

Magee, D., Zhu, Y., Ratnalingam, R., Gardner, P., Kessel, D. et al. (2007). An augmented reality simulator for ultrasound guided needle placement training. Medical & Biological Engineering & Computing, 45(10): 957–967.

Meltsner, M., Ferrier, N. J. and Thomadsen, B. (2007). Observations on rotating needle insertions using a brachytherapy robot. Physics in Medicine and Biology, 52(19): 6027.

Muntener, M., Patriciu, A., Petrisor, D., Mazilu, D., Bagga, H. et al. (2006). Magnetic resonance imaging compatible robotic system for fully automated brachytherapy seed placement. Urology, 68(6): 1313–1317.

Nath, S., Chen, Z., Yue, N., Trumpore, S., Peschel, R. et al. (2000). Dosimetric effects of needle divergence in prostate seed implant using 125i and 103pd radioactive seeds. Medical Physics, 27(5): 1058–1066.

Patil, S. et al. (2014). Needle steering in 3D via rapid replanning. IEEE Transactions on Robotics, 30(4): 853–864.

Patriciu, A., Petrisor, D., Muntener, M., Mazilu, D., Schar, M. et al. (2007). Automatic brachytherapy seed placement under mri guidance. IEEE Transactions on Biomedical Engineering, 54(8): 1499–1506.

Phee, L., Yuen, J., Xiao, D., Chan, C. F., Ho, H. et al. (2006). Ultrasound guided robotic biopsy of the prostate. International Journal of Humanoid Robotics, 3(04): 463–483.

Podder, T., Beaulieu, L., Caldwell, B., Cormack, R., Crass, J. et al. (2014). AAPM and GEC-ESTRO guidelines for image-guided robotic brachytherapy: Report of task group 192. Medical Physics, 41(10): 101501.

Rossa, C., Fong, J., Usmani, N., Sloboda, R., Tavakoli, M. et al. (2016). Multi-actuator haptic feedback on the wrist for needle steering guidance in brachytherapy. IEEE Robotics and Automation Letters, 99: 1–1.

Rossa, C., Khadem, M., Sloboda, R., Usmani, N., Tavakoli, M. et al. (2016). Adaptive quasi-static modelling of needle deflection during steering in soft tissue. IEEE Robotics and Automation Letters, 1(2): 916–923.

Rossa, C., Usmani, N., Sloboda, R. and Tavakoli, M. (2016). A handheld assistant for semi-automated percutaneous needle steering. IEEE Transactions on Biomedical Engineering In Press.

Salcudean, S., Prananta, T., Morris, W. and Spadinger, I. (2008). A robotic needle guide for prostate brachytherapy. In Robotics and Automation, 2008. ICRA 2008. IEEE International Conference on, pp. 2975–2981.

Salembier, C., Lavagnini, P., Nickers, P., Mangili, P., Rijnders, A. et al. (2007). Tumour and target volumes in permanent prostate brachytherapy: A supplement to the estro/eau/eortc recommendations on prostate brachytherapy. Radiotherapy and Oncology, 83(1): 3–10.

Schneider, C., Okamura, A. and Fichtinger, G. (2004). A robotic system for transrectal needle insertion into the prostate with integrated ultrasound. In IEEE International Conference on Robotics and Automation, 1: 365–370.

Seifabadi, R., Song, S., Krieger, A., Cho, N., Tokuda, J. et al. (2012). Robotic system for mri-guided prostate biopsy: Feasibility of teleoperated needle insertion and ex vivo phantom study. International Journal of Computer Assisted Radiology and Surgery, 7(2): 181–190.

Sloboda, R., Usmani, N., Pedersen, J., Murtha, A., Pervez, N. et al. (2010). Time course of prostatic edema post permanent seed implant determined by magnetic resonance imaging. Brachytherapy, 9(4): 354–361.

Taschereau, R., Pouliot, J., Roy, J. and Tremblay, D. (2000). Seed misplacement and stabilizing needles in transperineal permanent prostate implants. Radiotherapy and Oncology, 55(1): 59–63.

Usmani, N., Chng, N., Spadinger, I. and Morris, W. J. (2011). Lack of significant intraprostatic migration of stranded iodine 125 sources in prostate brachytherapy implants. Brachytherapy, 10(4): 275–285.

Waine, M., Rossa, C., Sloboda, R., Usmani, N. and Tavakoli, M. (2016). Needle tracking and deflection prediction for robot-assisted needle insertion using 2d ultrasound images. Journal of Medical Robotics Research, 01(01): 1640001.

Webster, R. J., Kim, J. S., Cowan, N. J., Chirikjian, G. S., Okamura, A. M. et al. (2006). Nonholonomic modeling of needle steering. The International Journal of Robotics Research, 25(5-6): 509–525.

Wei, Z., Wan, G., Gardi, L., Mills, G. and Downey, D. (2004). Robot-assisted 3D-TRUS guided prostate brachytherapy: System integration and validation. Medical Physics, 31(3): 539–548.

Wei, Z., Gardi, L., Downey, D. and Fenster, A. (2006). Automated localization of implanted seeds in 3d trus images used for prostate brachytherapy. Medical Physics, 33(7): 2404–2417.

Yu, Y., Podder, T., Zhang, Y., Ng, W. -S., Misic, V. et al. (2006). Robot-assisted prostate brachytherapy. In Medical Image Computing and Computer-Assisted Intervention, pp. 41–49.

Chapter 3

Team Cognition Assessment
From Concept to Practice

Bin Zheng

1. Team and Team Composition

Performing complex tasks requires more than one individual and focus to work in a cooperative environment. Increasing physical strengths in a team gives human operators the required muscles for handling tough task loads. More importantly, in today's work environment, the increasing cognitive capacity within a team provides the necessary power to deal with challenges brought by the compound tasks performed in an erratic environment (Salas et al., 2008). Winning a battle in an unfamiliar battlefield, or flying an airplane through unpredictable weather conditions, or saving lives in the operating room and the emergent room, human operators need to work in teams. Thus, we can have increasing physical and mental capacities for monitoring multiple channels of information, making appropriate decisions in a short time, and delivering multifaceted and synchronized actions to fulfill the task goal.

To build an effective team, a group of human individuals should 'interact dynamically, interdependently, and adaptively toward a common and valued goal, objective or mission, who have each been assigned specific roles or functions to perform, and have limited membership life span' (Salas et al., 1992). In the above definition, three key elements stand out that distinguish a *team* from a *group* of

Associate Professor in Surgery, Endowed Research Chair, Director, Surgical Simulation Research Lab, Faculty of Medicine & Dentistry, University of Alberta, 162 Heritage Medical Research Centre, 11207 – 87 Ave NW, Edmonton, AB, Canada T6G 2S2.
Email: bin.zheng@ualberta.ca

people, i.e., common goal, specific role, and interaction. Specifically, every individual in a team is assigned to a specific task to perform, and the role and responsibility of each team member should be known to all members in the team. They are working towards a common goal through team interaction, from taking and sharing information to delivering actions collaboratively. To work effectively together, team members must possess a specific set of knowledge for the team task; knowledge of their own and teammates' responsibilities, and 'a positive disposition toward working in a team' (Baker et al., 2006). The shared knowledge among team members is also referred to as the *team cognition*, covering the mutual understanding of how to collect and share information, manage available resources, support decision-making, and deliver a set of appropriate actions in responding to the changing environment (Cannon-Bowers and Salas, 2001).

A typical teamwork scenario is the cockpit of commercial airlines. There are always at least two pilots, and often there are three. The captain is the team leader who makes major decisions and leads the crew to ensure the safety of all passengers during normal or crisis moments. The flying of the airplane is shared between the captain and the first officer who normally sits on the right side of the cockpit. The first officer has been trained as the captain. They are sharing task loads, information, and decisions during the flight to keep flying error to a minimum (Martins, 2016). In some airlines, especially in old planes, a flight engineer is required, who has the responsibility of monitoring the airplane's instruments and calculating figures such as power setting, takeoff and landing speed, and fuel usage. In newer airlines, most of this work is done by computers on board. The flying team within the cockpit requires extensive team training to enable them to deal with crisis situations (Reynolds and Blickensderfer, 2009; Martins, 2016).

Another place we see teamwork in daily practice is the operating room for surgical care. Performing a life-saving surgical procedure requires a surgical team comprised of surgeons, anesthesiologists, and nurses (Zheng et al., 2012; Göras et al., 2019).

The primary surgeon is the person taking the lead role for the surgical team. The surgeon decides on the location of the incision, the approach to the surgical site, the range of tissue dissection, the volume of organ recession, and means of wound closure. The primary surgeon in the operating room often takes responsibility to ensure that the operation goes smoothly with minimal complications (Carthey et al., 2001). The surgeon's works are regularly assisted by two to four surgeons with different levels in surgical training. The assistant surgeons are often the clinical fellows and residents who are learning to perform the procedures by exposure to the surgical site and providing direct support to the primary surgeon. The primary surgeon needs to direct nurses and assistants in every step of the procedure while working closely with the anesthesiologist to manage a stable and safe patient condition throughout the procedure (Paige, 2010).

The surgeon's responsibility on patient safety is shared by the anesthesiologist in the team (Cooper, 2018). The anesthesiologists start their work for the patient before other surgical team members; they manage the patient's level of consciousness during the procedure to suppress the pain of the patient. The anesthesiologist has the principal responsibility to save patients' lives during various critical moments when the vital signs of patients show severe problems during surgery (Cooper and Gaba, 2002).

Nurses working inside the operating room are specially trained regularly (Alfredsdottir and Bjornsdottir, 2008). When they are assigned to work within the sterile field they are called scrub nurses. Their roles are to prepare and deliver surgical instruments to the surgeon, support the surgeon during the operation, and take care of the patient before and after entering the operating room. Circulating nurses work outside the sterile field in the operating room. They inspect surgical equipment and supplies, communicate with other medical staff outside the operating room, and support surgeons and anesthesiologists when needed in the operating room (Mitchell et al., 2011).

In a busy operating room, above core team members are assisted by surgical technicians, physician assistants, nurse anesthetists, and nursing students. Besides, these four categories of core team members, other health professionals may be called to the operating room to provide specific support to patients, including radiologists, cardiologists, endoscopists, ultrasound technicians, and industry representatives.

Starting in 2009, we study the size and composition of surgical teams from seven hospitals located in the United States, Canada, and China. Results have been published separately in the Journal of American Surgery (Cassera et al., 2009), the Canadian Journal of Surgery (Zheng et al., 2012), and the Surgical Endoscopy (He et al., 2014). Below we combine data from these three separate studies and give a global description on the size and composition of the surgical team. A total of 3160 surgical cases are reviewed in Table 1 and Table 2.

Table 1. Patients demographics, Procedure time, and Surgical teams from 3160 cases.

	Patient age (year)	ASA*	Procedure time (min)	Team size	Surgeon	Anesthetists	Nurse	Other**
Mean	52.4	2.1	89.4	7.5	2.7	1.3	2.8	0.2
Std. Dev.	17.1	0.7	57.0	1.2	0.7	0.3	0.5	0.4
Minimum	14	1	10	4	1	1	1	0
Maximum	94	4	535	14	6	4	7	5

Note: *ASA American Society of Anesthesiologists, **Others include radiologists, cardiologists, endoscopists, ultrasound technicians, and industry representatives.

Table 2. Surgical team composition by specialty.

#	Surgeons			Anesthesiologists			Nurses			Others		
	Freq	%	Cumulative	Freq	%	Cumulative	Freq	%	Cumulative	Freq	%	Cumulative
1	249	7.89	7.89	1711	54.16	54.16	11	0.36	0.36	1919	60.72	60.72
2	1592	50.37	58.26	1352	42.77	96.93	281	8.89	9.25	1062	33.62	94.34
3	1124	35.56	93.82	86	2.72	99.65	1480	46.84	56.09	141	4.47	98.81
4	167	5.27	99.09	11	0.35	100	889	28.12	84.21	27	0.84	99.65
5	20	0.63	99.72				360	11.4	95.61	11	0.35	100.00
6	9	0.28	100				95	3.01	98.62			
7							44	1.38	100			
Total	3160			3160			3160			3160		

A study on surgical team composition shows that, for routine general surgical procedures taking about 90 minutes, a surgical team contains as many as eight people on average, ranging from four to fourteen people (Table 1). Among them, one anesthetist and two nurses are a basic and requirement. However, the number of surgeons varied depending on different surgical procedures. Only 8% of procedures were performed by one surgeon; the majority were performed by two (50%) and three (36%) surgeons (Table 2).

Unlike the aviation crew, the surgical teams are not stable throughout the entire surgical procedure; members in a surgical team (such as nurses) leave and re-enter the team for various reasons, mainly regulated by the duty hour for different team members. The majority of surgical procedures involve three (47%) and four (28%) nurses. Making collaboration worse, the change of members within a surgical team is often unnoticed to the team leader. Planned actions with the previous team member(s) may not pass to the succeeding member(s). Therefore, increasing team size, complex composition, and unstable team formation are detrimental factors to the communication and collaboration of surgical teams (Carthey et al., 2001; Cassera et al., 2009; Zheng et al., 2012; He et al., 2014). Surgical time was prolonged as a result of a large team size in the operating room (Cassera et al., 2009; Zheng et al., 2012; He et al., 2014).

2. Team Cognition

We have an urgent need to train surgical teams after knowing the complicity of team composition and its detrimental impact on task performance (Birch et al., 2007; Paige, 2010; Fernandez et al., 2017). In such a dynamic work environment, surgical teams required team practice to develop team cognition (shared knowledge) among team members.

As illustrated in Fig. 1, team cognition is built based on individual cognition. Each individual has a limited capacity in processing information from the environment. In performing a daily task, we all maintain a comfort zone in processing information received from the environment. Taking driving as an example, information about the road, direction, and weather, etc., are taken by the driver through his/her visual channel. At the same time, the driver's control on the car is delivered by hands on the wheels and feet on the gas and brake pedals. Under routine driving, the driver's eye scanning and car manipulation are autonomous, meaning the

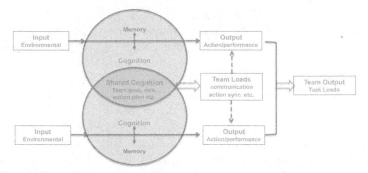

Figure 1. Conceptual model of information processing in individual and team. A portion of cognitive resources in each individual is used for coordinating team task.

volume of information processing is maintained in the comfort zone of his/her mind; the driver does not need to consciously focus on these tasks.

When it comes to crisis moments, such as driving in heavy rainfalls or passing a semi-truck, the driver will need to request all his/her attention and manipulate the vehicle deliberately. The volume of information processed by the driver in these difficult situations is dramatically increased, consuming a portion of spare mental resources in the mind. If any other new situation emerges such as the appearance of a bike rider on the road site, then the driver may have an inadequate cognitive capacity to deal with this sudden situation promptly. The consequence of overloading a driver's cognitive process can be quite a catastrophe.

Now we look into the information processing model in a team setting. Considering the driving tasks were performed by two operators, navigation and visual surveillance on the road can be performed by the assistant sitting beside a driver. When they are encountering some dangerous situations, they may have increased mental powers to deal with the problem and maintain safe driving. Theoretically, the total capacity for processing information related to the tasks will be double. However, in reality, a certain amount of mental resources from each team member will be used for communication and coordinating the actions between team partners. Therefore, the total mental resource of a team used for processing information related to the team tasks will be less than twofold. We name the portion of resource for sharing information and coordinating actions *team-related loads* and the portion of resource for focusing on team tasks the *task-related loads*.

In a newly-formed team, the team-related loads will be high because each team member has a strong desire for communication; they need to understand the partner's skill set, expectation and action plan to the team goal (Cannon-Bowers and Salas, 2001; Baker et al., 2006; Savelsbergh et al., 2012). Members in any new team will need time to practice together before they can develop a mutual understanding on how to communicate information and synchronize their movements. Once the team has received sufficient practice over time, the team-related loads can be decreased as they have developed a shared knowledge (team cognition) towards the team goal and the portion of mental resources for task-loads can be increased (Cannon-Bowers and Salas, 2001; Baker et al., 2006). We refer to a well-trained team as a *mature team*. A mature team can be silent since each team member has adequate knowledge on how to use individual skills to support each other to achieve a common team goal.

A good team training protocol is the one helping team members to reduce the amount of team-related loads yet increases the resources to the task-related load. In a certain type of surgical procedure, such as Minimally Invasive Surgery (MIS), the collaboration between the primary surgeon and the assistant is even more important because the visual presentation on surgical sites is controlled by the assistant who controls the camera, the laparoscope (Aggarwal et al., 2004). Therefore, surgical team training is critical for the success of any MIS procedure (Birch et al., 2009).

3. Team Cognition Assessment

One key element of building any team training program is to develop a reliable assessment instrument for measuring team cognition (Salas et al., 1992; Stout et al., 1999; Gisick et al., 2018). Traditionally, such an assessment is subjective, based on observation and judgement by the senior team observers (Cooke et al., 2000). Recently, a few new methodologies have been developed to objectively measure team cognition based on team collaborative behaviors. We are discussing these technologies for the assessment of team cognition in this chapter.

3.1 Observation-based Assessment

During team practices, behaviors of team collaboration and communication were classified and rated using paper-based assessment forms, such as the Oxford Nontechnical Skills (NOTECHS) for aviation pilots (Flin et al., 2003), the Multisource Feedback (MSF) for military (Atkins and Wood, 2002). Moving to the operating room for a surgical team, we have more specific tools like the Observational Teamwork Assessment for Surgery (OTAS) (Undre et al., 2007), the Nontechnical Skills (NOTSS) for surgical team (Yule et al., 2009), and the Anesthetists' Non-technical Skills (ANTS) for anesthetists (Fletcher et al., 2003).

The Oxford Non-Technical Skills (NOTECHS) scale was developed from an aviation instrument for assessment of teamwork and communication skills in the civil airline cockpit (Flin et al., 2003). The observation was done in four dimensions which were necessary for building a strong team: leadership and management, teamwork and cooperation, problem-solving and decision-making, and situation awareness. Within each dimension, there are 3 to 5 subteam modifiers. Team performance on each of these 16 subteam modifiers is scored on a scale of 1 to 4 (below standard −1, basic standard −2, standard = 3, excellent = 4), making a total score of 64 for the best team. Since its introduction in 2003, the NOTECHS evaluation system has been used in aviation and latterly in healthcare (Yule et al., 2006). Evidence for reliability and validity tests have been reported since then (Sevdalis et al., 2008; Mishra et al., 2009).

To apply the NOTECHS idea to the surgical theatre, Observational Teamwork Assessment for Surgery (OTAS) was developed in 2007 (Undre et al., 2007) and reported with rich evidence on reliability and validity tests (Sevdalis et al., 2009). The OTAS measures surgical team quality on five domains, communication, coordination, leadership, monitoring, and cooperation. The assessment can be performed before, during and after an operation. The OTAS rates performance of the surgeons, anesthetists, and nurses separately, covering all three core team members in any surgical team.

The abovementioned instruments for team assessment are observation-based, and often are straightforward and easy for execution. However, the restriction on these assessment tools is obvious. By inspecting team performance, the observer will rate the quality of team behaviors over a Likert scale on each domain and calculate a total score for team performance (Healey et al., 2004). Each form needs to have a validity test before it can be accepted by the research community (Sevdalis et al., 2009). Here, we need to make sure the content of the assessment form matching the team setting (content validity), the form can separate the skill levels among different teams (construct validity).

To control personal bias, observers need to undertake a long and structured training phase before team assessment. During the assessment, observers are required to maintain their focus on the team performance and capture behaviors of team members throughout the entire procedure. This may raise a problem in the surgical theatre. The surgical procedures are varied in length. Considering the mean surgical time can be over two hours; it is difficult for our observers to avoid developing fatigue and introducing personal bias after a long engagement in the assessment (Carthey et al., 2001). Quite often in true practice, the observations are aided by the video recording technique (Zheng and Swanström, 2009; Zheng et al., 2009). Ideally, we should find ways to assess team cognition directly rather than displayed behaviors.

One step towards the direct measurement of team cognition is by content analysis, which can be fulfilled using the *think-aloud* and the *multiple-choice questions* techniques (Cannon-Bowers et al., 1990). Specifically, we can compare the similarity of team knowledge among team members using cluster analysis, multidimensional scaling, pathfinder, and concept mapping.

Table 3 below shows an example of using multiple-choice questions test for assessing shared cognition among team members. Three emergency physicians provide answers to ten questions regarding their tactics task, each question has four (a, b, c, d) alternative choices.

Table 3. Hypothetic tests on team cognition using multiple choice questions.

Question	Correct answer	Subject A's response	Subject B's response	Subject C's response
1	C	c	c	c
2	D	d	d	d
3	B	b	b	b
4	A	d	d	d
5	B	c	c	c
6	D	b	d	d
7	B	d	b	b
8	A	d	c	a
9	D	a	b	d
10	A	c	d	b
Mean	(53%)	40%	50%	70%

On average, three physicians in the emergency medical team provided 53% correct answers to all questions; however, they only unanimously answered three questions (Q1, Q2, Q3) correctly. Interestingly, we found they all provide the same wrong answers in Q4 and Q5. This means, this team shared 50% of their team knowledge, but only 30% was correct, the other 20% shared knowledge was wrong. Further examining team members' answers, we found that two physicians provided correct answers to Q6 and Q7; one physician correctly answered Q8 and Q9); and for Q10, none of the team members gave correct answers. It seems that there were 70% chances the team might fail the team task due to wrong cognition and the incorrect actions they took.

Results presented by multiple-choice questions give us a good opportunity to visualize discrepancy and conflicts among team members and help us to comprehend where we can train team members to enhance their shared knowledge. Specifically, similar responses on Q4 and Q5 indicate the consensus, agreement, or overlap of shared knowledge among team members. Once we correct them with the right answers in these two situations, we can increase their shared knowledge to 50%.

Referring back to Table 1, we can make effects to teach Subject A on Q6 and Q7; improvement of Subject A on these two questions will increase shared knowledge of the team to 70%. Three physicians had a lower correction rate on Q8, Q9 and Q10; however, we noticed Subject C gave two correct answers to these three questions. The overall high score achieved by the Subject C indicates that he/she may have more experience in dealing with difficult situations. In reality, we may ask Subject A and Subject B to trust Subject C's judgments to increase the chance of taking the right actions in the emergent cases. All wrong answers to Q10 are warning us that special training on this emergent situation is needed before we can comfortably send the team to the emergency room.

In brief, by conducting detailed analysis on the answers to multiple choice questions, we are able to identify the problems in a team, find ways to solve the problems, and build up shared knowledge promptly and appropriately among team members (Graesser et al., 2018).

3.2 Behavioral-based Assessment

Over the past decades, our team performed teamwork assessments in the surgical context using behavior markers displayed by the surgical team. We started by setting up a complex laparoscopic task that required two instruments working together. The laparoscopic procedure is a robotic and image-guided surgery performed in the abdominal cavity of patients. Surgeons in the laparoscopic procedure are required to complete the surgical tasks using long-shafted instruments while watching images of the surgical sites which are captured by an endoscope and displayed on a television monitor.

In the simulation environment, operators were recruited to reach and grasp a surgical thread, pull it up from a piece of synthetic soft tissue for a short distance using a laparoscopic grasper in one hand, and then cutting it with a pair of endoscopic scissors held in the other hand. The task can be performed by one surgeon bi-manually, or by two surgeons uni-manually working side by side in a dyad team (Zheng et al., 2005; Zheng et al., 2007). All subjects were right-handed, thus for the given task, subjects in the individual bimanual group held the grasper in the left and the scissors in the right hand. In the dyad team group, each instrument was held in the team members' preferred hand.

Results showed that total task time was shorter in the dyad team compared to the individual group. The contribution to the shorter total task time in the dyad team group was primarily due to the smaller delay between the start of the grasper and scissor movements rather than the difference in the movement speeds of these two instruments (Zheng et al., 2005). When the identical surgical task was performed by the individual bimanually, the scissors started to move to the surgical site quite late, often after the grasper completed grasping the target. Whereas in the dyad team, the scissors regularly were moved up nearby the cutting site before the grasper held the thread completely (Zheng et al., 2007). Since no verbal communication was allowed in the dyad team group, the early scissors movements performed by the scissor holders were not following the instructions of the grasper holder. Rather, the performance of this proactive movement was driven by the intention of a team member who clearly understood the team goal, roles of each member in the team, and was willing to collaborate to fulfill the team task as quickly as possible. We call this type of proactive action within a team without being instructed by another team member *anticipatory movement* (Zheng et al., 2007).

In contrast, individual operators were overloaded by challenges in laparoscopic tasks. A single operator can only mentally focus on the action performed in one hand before shifting attention to another hand. Each human operator has a limited capacity to process information loaded to his/her cognitive pathway. In case

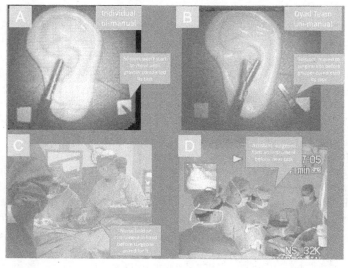

Figure 2. Anticipatory movements performed in the laboratory and clinical settings. In a laboratory setting (A, B), the scissor holder in the dyad team started movement before his team partner grasped the thread ready for cutting (B); in operating room (C, D), the scrubbing nurse (C), and the assistant surgeon (D) were recorded to perform anticipatory movements during surgical procedures.

of performing tasks so challenging as in laparoscopic surgery, we failed to see individual operators concurrently handle tasks with both hands (Zheng et al., 2007). However, the increasing mental power brought by the team members enables them to perform multiple tasks simultaneously. This phenomenon was more visible when giving them time to practice. Giving time to practice together, scissors moving up early to the surgical site were observed in all trials (Carthey et al., 2001; Zheng et al., 2007; Mitchell et al., 2011; Hicks and Petrosoniak, 2018). Therefore, the performance of antiparty movement has been recognized as a valuable behavioral marker for collaborative team works; and was observed in retrospect in many clinical settings.

In 2006, we conducted a field study in the operating room. We observed over 150 hours of surgical procedures on 59 laparoscopic cases performed by a group of surgical teams assisted by intermediate and experienced operating (scrubbing) nurses (Zheng et al., 2009). By checking the nurse's actions every minute during the operation, we noticed that their surgery-related activities can be roughly divided into two categories, with or without following a verbal instruction from the surgeon. As you can imagine, a majority of nurse activities followed the instructions of surgeons; yet there were occasions the supportive actions were performed by the nurses without following an instruction. The nurse anticipated the need of the surgeon and prepared the action ahead of time (Zheng et al., 2009). For example, a nurse readied a pair of scissors ready when she saw the surgeon tying a knot; once the surgeon withdrew the needle driver from the suturing site, indicating the completion of knot tying, the nurse delivered the scissor to the surgeon's hand without the need for a verbal instruction.

The ability to perform anticipatory movements is regulated by a nurse's experience in participating in surgical cases (Zheng et al., 2009; Mitchell et al., 2011). We found that intermediate and experienced nurses performed a similar amount of anticipatory movements in relatively easy and routine cases. When supporting complicated cases involving many on-site decisions, experienced nurses had the ability to anticipate surgeons' actions; they performed significantly more anticipatory movements than the intermediate nurses (Zheng et al., 2009).

Aided with video cameras, we recorded inter-person interaction between surgeons during the laparoscopic procedure (Zheng et al., 2008; Zheng and Swanström, 2009). When the surgical team was newly formed, with the induction of a surgical fellow to the surgical unit by rotation, small amounts of anticipatory movements were observed compared to those performed by the original surgical team which had handled more surgery cases together. As the surgical team became mature, members in the team fully understood the role, abilities, and personal preference of their teammates; they could predict each other's expected actions and performed collaborative tasks without the need for verbal communication (Zheng and Swanström, 2009). Increasing evidence has suggested that anticipatory movement is a valuable behavioral marker for team collaboration.

Having said that, we are aware that collaborative behaviors such as anticipatory movements are not always aligned with the level of shared cognition within a team. In some situations, such as a team with active verbal communication, members in a team may not have chances to perform an increasing number of anticipatory movements even though they possess a high level of shared team knowledge (Carthey et al., 2001; Mitchell et al., 2011; Nurok et al., 2011). We still need to find direct ways to measure team cognition in the minds of team members. The advancement in applying dual eye-tracking in the team study opens a possibility to achieve this goal.

3.3 Dual Eye-tracking for Team Cognition Assessment

Eye-tracking is the process of measuring the point of gaze when a person is taking in information with his/her eyes. To achieve this, we need to use a device called eye-tracker. Most eye-trackers used today are non-intrusive; the reflecting lights on the surface of eyes can be recorded and used for calculating the point of gaze.

According to the eye-mind hypothesis proposed by Just and Carpenter, what is fixated in the eyes is coupled with what is processed in mind (Just and Carpenter, 1976). However, this hypothesis was challenged later by pieces of evidence in human observation. We are occasionally thinking about something that we are not looking at. The existence of this type of *covert attention* warns us not to simply use fixation points for indicating fixation (Posner, 1980; Mulckhuyse and Theeuwes, 2010; Posner, 2016). However, the fast-shifting gaze from one location to another (saccade eye movement) can be safely interpreted as the relocation of attention. Also, when the gaze is connecting to actions in hand, covert attention is difficult to happen. Such as in surgery where surgeons' visions are actively engaged to the task for guided manipulation, when the points of eyes are highly aligned with the points of mental focus. Therefore, if we can record each team members' gaze points over a surgical procedure, our analyses on the similarity of their gaze points over time (the trajectories of the gazes) should shed light on the level of shared cognition among team members towards the team task.

One barrier before us is how to record eye movements from multiple users. Typically in studying human eye movement, one eye-tracker unit is required for every operator; it is difficult to track multiple eye movements with current technology. In 2011, our team started to develop dual eye-tracking technology (Khan et al., 2012; Atkins et al., 2013). As shown in Fig. 3, two separate Tobii eye-trackers were used to track two surgeons' eye movements while they were performing a laparoscopic team task. Two eye-tracking signals were obtained and superimposed over the same surgical video with identical time frames.

With this innovative technology, we were able to examine the similarity of two surgeons and test whether a mature team displays more collaborative eye behavior than the immature team. Our initial

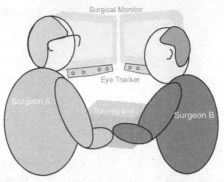

Figure 3. Recording dual eye-tracking from two surgeons during a laparoscopic surgery.

attempt in checking the similarity between two eye motions was quite straight forward. At any time frame, we calculated the Euclidean distance between two gaze points (Tien et al., 2013). If the distance fell within 200 pixels, we recorded a moment of *gaze overlap* (Fig. 4). In an early study of comparing operators' gaze points while performing a task versus just watching the task video, we found gaze overlap among operators is at a higher level than when they watched the video.

The 200-pixel criterion for gaze overlap was arbitrated as it could be translated to a 1-cm distance in the monitor used for the study at that time. Within this distance, surgeons were supposed to be looking at the same anatomic structure. We can also use the visual angle to set up the threshold for gaze overlap. In Khan and Zheng's study on surgeons, we defined the 3° visual field for the area of common interest (Khan et al., 2012). We calculated the presentation of gaze overlap over entire surgical procedures and compared the difference between mature and immature surgical teams. The results were encouraging; two operators with rich surgical experiences achieved 55% gaze overlap which is significantly higher than the 43% gaze overlap recorded from two operators with poor surgical experiences.

Checking gaze overlap at any given time is simple and fragile to describe complex and dynamic patterns of eye movement. Before introducing a more complicated algorithm, we need to know the nature of the eye-tracking signal. The eye-tracking data is comprised of a series of position data (X,Y) calculated over two coordinates, one for the camera and the other for the world. Once calibrated well, these two coordinates should be aligned, i.e., the position data should point to where a person looks at. The calculations on the Euclidean

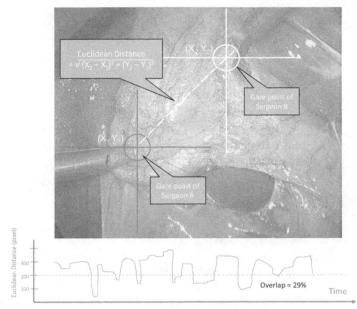

Figure 4. Calculation of gaze overlap between two surgeons in laparoscopic surgery. Euclidean distances between two gaze points were calculated at any time frame and displayed over entire surgical time. Distance less than 200 pixel was recorded as the *gaze overlap.*

distance or the visual angles are all based on the position data of gazes.

Each team member may have a different visual scanning approach to the environment. We use a hypothetical model to explain the situation. Considering there are five spots in the visual field to inspect. Operator X fixates these spots from A to E; Operator Y also fixated on all these spots but with a different starting point, for example from C to E then back to A. The scanning trajectories of these two operators arc displayed in Fig. 5. Purely checking overlap among them will yield an unexpectedly low rate (Fig. 5, lower left) that may not represent the true nature of these two operators in terms of their shared knowledge towards team tasks. In reality, both operators inspect the identical hot spots and they may already generate similar task plans even though they are dis-synchronizing their visual scan paths. The question for us is how to correct the problem in the current method for calculating gaze overlap and finding an improved but robust way to describe members' shared knowledge displayed in their eye scanpaths.

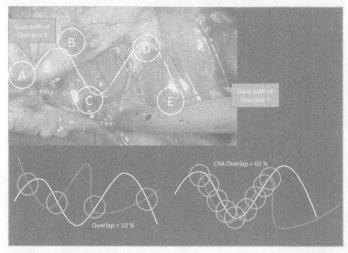

Figure 5. Illustration of applying the cross-recurrence analysis to eye scanpath to improve the calculation of gaze overlap between two surgeons.

The answers can be found in the *cross-recurrence analysis* (Hajari et al., 2016). Using this algorithm, the scanpath of Operator Y is shifted along the time axis until maximum overlapping is achieved with Operator X (Fig. 5, lower right). A typical CRA report includes phase lag as well as maximum overlap. By taking both spatial and tempera features of the eye scanpath into its calculation, CAR yields a better mathematical outcome in checking the similarity among different scanpaths, which enhances our ability to describe shared cognition among team members.

In summary, team cognition is the foundation for team performance (Cannon-Bowers et al., 1990; Salas et al., 2008). When each member in a team understands the team goal, roles of each other in the team, and ways to collaborate for the goal, they will anticipate actions and fulfill the task goal in a fast and effective manner. Besides those observational methods, we can measure team cognition using behavioral markers captured by video and eye-tracking. Videos record an increasing number of anticipatory movements in the mature teams, whereas eye-tracking discovered a higher rate of gaze overlap among members in the mature teams. Although most pieces of evidence presented in this chapter were collected from healthcare, results can be applied to other industries that are heavily dependent on team work. As new tracking technologies will be available for studying team collaboration, we anticipate more behavioral pieces of evidence will be produced towards our goal of assessing team performance and cognition with objective methods.

References

Aggarwal, R., Moorthy, K. and Darzi, A. (2004). Laparoscopic skills training and assessment. Br. J. Surg., 91(12): 1549–1558.

Alfredsdottir, H. and Bjornsdottir, K. (2008). Nursing and patient safety in the operating room. J. Adv. Nurs., 61(1): 29–37.

Atkins, M. S., Tien, G., Khan, R. S., Meneghetti, A., Zheng, B. et al. (2013). What do surgeons see: capturing and synchronizing eye gaze for surgery applications. Surg. Innov., 20(3): 241–248.

Atkins, P. W. B. and Wood, R. E. (2002). Self-versus others' ratings as predictors of assessment center ratings: Validation evidence for 360-degree feedback programs. Pers. Psychol., 55(4): 871–904.

Baker, D., Day, R. and Salas, E. (2006). Teamwork as an essential component of high-reliability organizations. Health Serv. Res., 41: 1576–1598.

Birch, D. W., Misra, M. and Farrokhyar, F. (2007). The feasibility of introducing advanced minimally invasive surgery into surgical practice. Can. J. Surg., 50(4): 256–260.

Birch, D. W., Bonjer, H. J., Crossley, C., Burnett, G., de Gara, C. et al. (2009). Canadian consensus conference on the development of training and practice standards in advanced minimally invasive surgery. Can. J. Surg., 52(4): 321–327.

Cannon-Bowers, J. A., Salas, E. and Converse, S. A. (1990). Cognitive psychology and team training: Training shared mental models and complex systems. Hum Factors Bullet, 33(12): 1–4.

Cannon-Bowers, J. A. and Salas, E. (2001). Reflections on shared cognition. J. Organ. Behav., 22(2): 195–202.

Carthey, J., de Leval, M. R. and Reason, J. T. (2001). The human factor in cardiac surgery: Errors and near misses in a high technology medical domain. Ann. Thorac. Surg., 72(1): 300–305.

Cassera, M. A., Zheng, B., Martinec, D. V., Dunst, C. M., Swanström, L. L. et al. (2009). Surgical time independently affected by surgical team size. Am. J. Surg., 198(2): 216–222.

Cooke, N. J., Salas, E., Cannon-Bowers, J. A. and Stout, R. J. (2000). Measuring team knowledge. Hum Factors, 42(1): 151–173.

Cooper, J. B. and Gaba, D. (2002). No myth: Anesthesia is a model for addressing patient safety. Anesthesiology, 97(6): 1335–1337.

Cooper, J. B. (2018). Critical role of the surgeon-anesthesiologist relationship for patient safety. Anesthesiology, 129(3): 402–405.

Fernandez, R., Shah, S., Rosenman, E. D., Kozlowski, S. W. J., Parker, S. H. et al. (2017). Developing team cognition: A role for simulation. Simul. Health., c12(2): 96–103.

Fletcher, G., Flin, R., McGeorge, P., Glavin, R., Maran, N. et al. (2003). Anaesthetists' Non-Technical Skills (ANTS): Evaluation of a behavioural marker system. Br. J. Anaesth., 90(5): 580–588.

Flin, R. H., Martin, L., Goeters, K. -M., Hormann, H. J. Amalberti, R. et al. (2003). Development of the NOTECHS (non-technical skills) system for assessing pilots' CRM skills. Hum Factors Aerosp Safety, 3(2): 97–119.

Gisick, L. M., Webster, K. L., Keebler, J. R., Lazzara, E. H., Fouquet, S. et al. (2018). Measuring shared mental models in healthcare. J. Patient Safety Risk Manag., 23(5): 207–219.

Göras, C., Olin, K., Unbeck, M., Pukk-Härenstam, K., Ehrenberg, A. et al. (2019). Tasks, multitasking and interruptions among the surgical team in an operating room: A prospective observational study. BMJ Open, 9(5): e026410.

Graesser, A. C., Fiore, S. M., Greiff, S., Andrews-Todd, J., Foltz, P. W. et al. (2018). Advancing the science of collaborative problem solving. Psychol. Sci. Public Interest, 19(2): 59–92.

Hajari, N., Cheng, I., Zheng, B. and Basu, A. (2016). Determining team cognition from delay analysis using cross recurrence plot. 2016 38th Annual International Conference of the IEEE Engineering in Medicine and Biology Society (EMBC).

He, W., Ni, S., Chen, G., Jiang, X., Zheng, B. et al. (2014). The composition of surgical teams in the operating room and its impact on surgical team performance in China. Surg. Endosc., 28(5): 1473–1478.

Healey, A. N., Undre, S. and Vincent, C. A. (2004). Developing observational measures of performance in surgical teams. Qual. Saf. Health Care, 13 Suppl 1: i33–40.

Hicks, C. and Petrosoniak, A. (2018). The human factor: Optimizing trauma team performance in dynamic clinical environments. Emerg. Med. Clin. North Am., 36(1): 1–17.

Just, M. A. and Carpenter, P. A. (1976). Eye fixations and cognitive processes. Cogn. Psychol., 8(4): 441–480.

Khan, R. S., Tien, G., Atkins, M. S., Zheng, B., Panton, O. N. et al. (2012). Analysis of eye gaze: Do novice surgeons look at the same location as expert surgeons during a laparoscopic operation? Surg. Endosc., 26(12): 3536–3540.

Martins, A. (2016). A review of important cognitive concepts in aviation. Aviation, 20: 65–84.

Mishra, A., Catchpole, K. and McCulloch, P. (2009). The Oxford NOTECHS system: Reliability and validity of a tool for measuring teamwork behaviour in the operating theatre. Qual. Saf. Health Care, 18(2): 104.

Mitchell, L., Flin, R., Yule, S., Mitchell, J., Coutts, K. et al. (2011). Thinking ahead of the surgeon. An interview study to identify scrub nurses' non-technical skills. Int. J. Nurs. Stud., 48(7): 818–828.

Mulckhuyse, M. and Theeuwes, J. (2010). Unconscious attentional orienting to exogenous cues: A review of the literature. Acta Psychol. (Amst.), 134(3): 299–309.

Nurok, M., Sundt, T. M. 3rd and Frankel, A. (2011). Teamwork and communication in the operating room: Relationship to discrete outcomes and research challenges. Anesthesiol. Clin., 29(1): 1–11.

Paige, J. T. (2010). Surgical team training: promoting high reliability with nontechnical skills. Surg. Clin. North Am., 90(3): 569–581.

Posner, M. I. (1980). Orienting of attention. Q. J. Exp. Psychol., 32(1): 3–25.

Posner, M. I. (2016). Orienting of attention: Then and now. Q. J. Exp. Psychol., (2006) 69(10): 1864–1875.

Reynolds, R. and Blickensderfer, E. (2009). Crew resource management and shared mental models: A proposal. J. Avia. Aerosp. Edu. Res., 19(1): 15–23.

Salas, E., Dickinson, T. L., Converse, S. A. and Tannenbaum, S. I. (1992). Toward an understanding of team performance and training. Teams: Their training and performance. Westport, CT, US, Ablex Publishing, 3–29.

Salas, E., Cooke, N. J. and Rosen, M. A. (2008). On teams, teamwork, and team performance: Discoveries and developments. Hum Factors, 50(3): 540–547.

Savelsbergh, C., Gevers, J. M. P., van der Heijden, B. I. J. M. and Poell, R. F. (2012). Team role stress: Relationships with team learning and performance in project teams. Group Organ. Manag., 37(1): 67–100.

Sevdalis, N., Davis, R., Koutantji, M., Undre, S., Darzi, A. et al. (2008). Reliability of a revised NOTECHS scale for use in surgical teams. Am. J. Surg., 196(2): 184–190.

Sevdalis, N., Lyons, M., Healey, A. N., Undre, S., Darzi, A. et al. (2009). Observational teamwork assessment for surgery: Construct validation with expert versus novice raters. Ann. Surg., 249(6): 1047–1051.

Stout, R. J., Cannon-Bowers, J. A., Salas, E. and Milanovich, D. M. (1999). Planning, shared mental models, and coordinated performance: An empirical link is established. Hum Factors, 41(1): 61–71.

Tien, G., Atkins, M. S., Jiang, X., Khan, R. S., Zheng, B. et al. (2013). Identifying eye gaze mismatch during laparoscopic surgery. Stud. Health Technol. Inform., 184: 453–457.

Undre, S., Sevdalis, N., Healey, A. N., Darzi, A., Vincent, C. A. et al. (2007). Observational teamwork assessment for surgery (OTAS): Refinement and application in urological surgery. World J. Surg., 31(7): 1373–1381.

Yule, S., Flin, R., Paterson-Brown, S. and Maran, N. (2006). Non-technical skills for surgeons in the operating room: A review of the literature. Surgery, 139(2): 140–149.

Yule, S., Rowley, D., Flin, R., Maran, N., Youngson, G. et al. (2009). Experience matters: Comparing novice and expert ratings of non-technical skills using the NOTSS system. ANZ J. Surg., 79(3): 154–160.

Zheng, B., Verjee, F., Lomax, A. and MacKenzie, C. L. (2005). Video analysis of endoscopic cutting task performed by one versus two operators. Surg. Endosc., 19(10): 1388–1395.

Zheng, B., Swanström, L. L. and MacKenzie, C. L. (2007). A laboratory study on anticipatory movement in laparoscopic surgery: A behavioral indicator for team collaboration. Surg. Endosc., 21(6): 935–940.

Zheng, B., Martinec, D. V., Cassera, M. A. and Swanström, L. L. (2008). A quantitative study of disruption in the operating room during laparoscopic antireflux surgery. Surg. Endosc., 22(10): 2171–2177.

Zheng, B. and Swanström, L. L. (2009). Video analysis of anticipatory movements performed by surgeons during laparoscopic procedures. Surg. Endosc., 23(7): 1494–1498.

Zheng, B., Taylor, M. D. and Swanström, L. L. (2009). An observational study of surgery-related activities between nurses and surgeons during laparoscopic surgery. Am. J. Surg., 197(4): 497–502.

Zheng, B., Panton, O. N. and Al-Tayeb, T. A. (2012). Operative length independently affected by surgical team size: Data from 2 Canadian hospitals. Can. J. Surg., 55(6): 371–376.

Chapter 4

How to Make Sure That Robot Behavior Is Human-Like

Vladik Kreinovich, Olga Kosheleva and *Laxman Bokati*

1. Introduction and Formulation of the Problem

Need for robots that look and act like humans: a brief reminder. In many practical applications, it is desirable to have robots that look and act like humans. For example, if we want to create a robot that takes care of small children, it is desirable to have a human-like robot, to utilize the children's natural affinity towards human beings and their natural fear of unusual creatures. Similarly, a medical robot that looks and acts like a human will hopefully help the patients to be somewhat more relaxed in an already stressful illness situation. A robot that takes care of older people will sound warmer if this robot is more human-like. And, of course, if something happens to a human operator (e.g., human driver, human pilot, etc.), it would be great for the robot to be able to fit into the control seat and take over.

What was the main challenge in the past decades. In the past, the main problem was to make a robot look and behave like a human.

What is the main problem now. Nowadays, we have robots that look and sometimes behave remarkably like humans – we have robotic TV announcers that are difficult to distinguish from the real ones, robotic performers, among others.

University of Texas at El Paso, 500 W. University, El Paso, TX 79968, USA.
Emails: vladik@utep.edu; olgak@utep.edu; lbokati@miners.utep.edu

So we face a different problem now – that excellent robots can be made. Their movements can be made optimal up to the minutest detail, and their decisions can be made optimal. As a result, while these robots may look like humans, they do not behave like humans – to be more precise, instead of behaving like us humans, the robots behave like idealized never-making-a-mistake superhuman beings.

Clarification: sometimes we do need superhuman robots. Robots with superhuman abilities are definitely needed in many applications – we need robots that can bravely go where humans cannot, that can explore space, rescue victims of earthquakes, repair nuclear reactors.

In many other applications, we need more human-like robots. Having superhuman ability in a humanoid robot defeats its very purpose – to look and behave like a human.

Of course, we want this robot to act like a very good human – e.g., we do not want a medical robot to make mistakes on purpose. However, in their movements, and appearance, we do not want them to be perfect, but we want them to be like us. Misdiagnosing a patient is a big no-no, but why not make a robot "accidentally" bump into a chair and move it slightly (as a human being would) if this will make this robot (and thus, this robot's advice) more acceptable to the patients.

What we do in this chapter. In this chapter, we provide algorithmic foundations that will (hopefully) help in designing such human-like robots.

2. Analysis of the Problem

Main idea. How can we simulate suboptimal human behavior? According to modern psychology, as discovered and emphasized by the Nobelists Herbert Simon and Daniel Kahneman (in his collaboration with Amos Tversky), this sub-optimality is mostly due to *bounded rationality* – i.e., to the fact that we humans have limited ability to process information; see, e.g., Kahneman, 2013.

Details. As a result, e.g., when optimizing, we do not exactly find *the* value of the parameters for which the objective function attains its largest value – instead:

- we first discretise the problem, by dividing the range of possible situations (i.e., of possible values of the parameters) into finitely many subranges, and

■ then, in each subrange, we select a typical situation, and we find the decision which is optimal for this typical situation; this decision will be used for all situations from this subrange.

This is how we deal with most real-life problems; for example:

■ in a big class, where an individual approach is not realistically possible, an instructor deals separately with A students (they need extra assignments), with C students (they need encouragement), and with potentially failing students (they need help);

■ a medical doctor diagnoses a patient, and then prescribes the medicine corresponding to this particular diagnosis and this particular group of patients, etc.

How should we select subranges? Once we apply the same solution to all the situations from a given subrange, our solutions become suboptimal. For some divisions into subranges, we may have, in some situations, a big deviation from optimality. For other divisions, the deviations are not that large.

It is reasonable to select a division which is *optimal* – in the sense that the resulting decisions are as close to the optimal one as possible for the given fixed number of subranges.

What we do in this chapter. In this chapter, we formulate the corresponding optimization problem – of selecting the optimal division into subranges – in precise terms, and provide a solution to this problem.

We will consider two possible cases:

■ the case when we do not know the probabilities of different situations; in this case, the natural way to gauge suboptimality is by the worst-case difference between the optimal and suboptimal values of the corresponding objective function, and

■ the case when we know the probabilities of different situations; in this case, it is more natural to gauge sub-optimality by the *average* difference between the optimal and suboptimal values.

Comment. Some of the corresponding mathematics will be similar to the one used in the book (Lorkowski and Kreinovich, 2018)

to describe human behavior. In particular, this book shows that the idea of bounded rationality explains why humans use imprecise ("fuzzy") natural-language terms when making decisions, and why fuzzy control – that takes these words into account – often performs better than a probabilistic approach; see, e.g., (Belohlavek et al., 2017; Klir and Yuan, 1995; Mendel, 2017; Nguyen et al., 2019; Novák et al., 1999; Zadeh, 1965). The explanation is that the fuzzy approach implicitly takes into account not only the probability of different alternatives, but also their utility.

The main difference of what we do in this chapter from the mathematical analysis presented in (Lorkowski and Kreinovich, 2018) is that there, our main goal was to *describe* human behavior, while here, the objective is to *recommend* (*prescribe*) the robot's behavior.

3. Case When We Do Not Know the Probabilities of Different Situations

General description of the control situation. To describe a situation, we need to describe the values of the quantities $x = (x_1, \ldots, x_n)$ that describe this situation. For example, the state of a doctor's patient can be described by the patient's body temperature, age, blood pressure, etc. The state of a student can be characterized by the student's grades on different assignments. The state of a mobile robot can be characterized by the coordinates describing its location and by the components of the velocity vector. If the robot has arms, we should also describe the angles between different parts of the robot's arm and the corresponding angular velocities.

Not all possible combinations of parameters are usually realistically possible. Let X denote the set of possible values of the tuples x.

To improve the situation, we can apply different controls. Control can also be characterized by the values of the corresponding parameters $u = (u_1, \ldots, u_m)$. For example, we can slow down or speed up the robot, change its direction, lift of lower its arm, etc.

We usually know the *objective function*, i.e., we know the gain $G(x, u)$ that we will get if we are in the state x, and we apply the control u. For example, if the goal is for a robot to reach the patient within a certain period of time t_0 (e.g., if the patient fell down), and the sooner the better, then $G(x, u)$ is the difference between

t_0 and the time that the robot in the original state x will take to reach the patient after applying the control u.

In general, our objective is to maximize this gain.

Ideal case: optimal control. In the ideal case, for each situation x, we should selected the *optimal* control $u^{\mathrm{opt}}(x)$, i.e., control for which the gain is the largest:

$$G\left(x, u^{\mathrm{opt}}(x)\right) = \max_u G(x, u). \tag{1}$$

Case of human-like behavior. In the case of human-like behavior, we divide the range X into subranges X_j. In each subrange, we select a typical representative situation $x^{(j)}$, and apply the control $u^{\mathrm{opt}}\left(x^{(j)}\right)$ to all situations from the subrange X_j.

How to describe the degree of sub-optimality. For each situation $x \in X_j$, the best we can do is to apply the control $u^{\mathrm{opt}}(x)$ which is optimal for this situation. Then, we will get the gain $G\left(x, u^{\mathrm{opt}}(x)\right)$. Instead, we get the gain $G\left(x, u^{\mathrm{opt}}\left(x^{(j)}\right)\right)$. The difference between these gains is equal to

$$\Delta G_j(x) = G\left(x, u^{\mathrm{opt}}(x)\right) - G\left(x, u^{\mathrm{opt}}\left(x^{(j)}\right)\right). \tag{2}$$

For a close-to-optimal control, the subranges are small, and all the situations within each subrange are close to each other, so

$$x = x^{(j)} + \Delta x, \tag{3}$$

for some small Δx, and, correspondingly,

$$u^{\mathrm{opt}}(x) = u^{\mathrm{opt}}\left(x^{(j)}\right) + \Delta u, \tag{4}$$

for some small Δu. We can therefore substitute the expression

$$u^{\mathrm{opt}}\left(x^{(j)}\right) = u^{\mathrm{opt}}(x) - \Delta u \tag{5}$$

into the formula (2):

$$\Delta G_j(x) = G\left(x, u^{\mathrm{opt}}(x)\right) - G\left(x, u^{\mathrm{opt}}(x) - \Delta u\right), \tag{6}$$

expand this expression in Taylor series, and keep the largest non-zero terms in this expansion. In general, we have

$$G\left(x, u^{\mathrm{opt}}(x) - \Delta u\right) = G\left(x, u^{\mathrm{opt}}(x)\right) - \sum_{i=1}^{m} \frac{\partial G(x, u)}{\partial u_i} \cdot \Delta u_i +$$

$$\frac{1}{2} \cdot \sum_{i=1}^{m} \sum_{j=1}^{m} \frac{\partial^2 G(x,u)}{\partial u_i \partial u_j} \cdot \Delta u_i \cdot \Delta u_j + \ldots \tag{7}$$

By definition (1) of the optimal control $u^{\text{opt}}(x)$, the function $G(x,u)$ attains its maximum at this control, so all the partial derivatives are equal to 0:

$$\frac{\partial G(x,u)}{\partial u_i} = 0, \tag{8}$$

thus

$$G\left(x, u^{\text{opt}}(x) - \Delta u\right) = G\left(x, u^{\text{opt}}(x)\right) + \frac{1}{2} \cdot \sum_{i=1}^{m} \sum_{j=1}^{m} \frac{\partial^2 G(x,u)}{\partial u_i \partial u_j} \cdot \Delta u_i \cdot \Delta u_j + \ldots$$

$$\tag{9}$$

Substituting this expression into the formula (6), we conclude that the main term in the difference (6) is quadratic:

$$\Delta G_j(x) = -\frac{1}{2} \cdot \sum_{i=1}^{m} \sum_{j=1}^{m} \frac{\partial^2 G(x,u)}{\partial u_i \partial u_j} \cdot \Delta u_i \cdot \Delta u_j. \tag{10}$$

Here,

$$\Delta u_i = \sum_{a=1}^{n} \frac{\partial u_i^{\text{opt}}(x)}{\partial x_a} \cdot \Delta x_a. \tag{11}$$

Thus, the formula (1) takes the form

$$\Delta G_j(x) = \sum_{a=1}^{n} \sum_{b=1}^{n} c_{ab,j} \cdot \Delta x_a \cdot \Delta x_b, \tag{12}$$

where we denoted

$$c_{ab,j} = c_{ab}\left(x^{(j)}\right)$$

and

$$c_{ab}(x) = \sum_{i=1}^{m} \sum_{j=1}^{m} \frac{\partial^2 G(x,u)}{\partial u_i \partial u_j}_{\big| u = u^{\text{opt}}(x)} \cdot \frac{\partial u_i^{\text{opt}}(x)}{\partial x_a} \cdot \frac{\partial u_j^{\text{opt}}(x)}{\partial x_b}. \tag{13}$$

The overall quality of division into subranges is described by the worst-case value of $\Delta G_j(x)$, i.e., by the value

$$\max_{j} \max_{x \in X_j} \Delta G_j(x). \tag{14}$$

We want to find the division into subranges for which the quantity (14) is the smallest possible.

Towards finding the optimal division into subranges. For each region X_j, let $v_j = \max\limits_{x \in X_j} \Delta G_j(x)$ be the largest value of $\Delta G(x)$ for all points $x \in X_j$. Then, it makes sense to assign, to this region, all the points x for which $\Delta G_j(x) \le v_j$ – adding these points to X_j will not increase the worst-case sub-optimality.

According to the formula (12), the value $\Delta G_j(x)$ is a quadratic function of x_a, so each region $\{x : \Delta G_j(x) \le v_j\}$ is an ellipsoid. Thus, in the first crude approximation, each subrange X_j is an ellipsoid whose axes are eigenvectors of the matrix a_{ij}. However, these subranges needs to fill the whole space, so we need to make them parallelepipeds with axes parallel to the axes of the ellipsoid – i.e., to the eigenvectors of the matrix a_{ij}.

If for some j_0, we have $v_{j_0} < \max\limits_{j} v_j$, then we can increase the subrange X_{j_0} and decrease the size (and thus, the values v_j) for other subranges, thus decreasing the value $\max\limits_{j} v_j$. Thus, in the optimal division into subranges, we should have all values v_j equal. Let us denote the common value of all these v_j by v.

What value v should we select? Suppose that we want to divide the whole range X into N subranges. At each point x, the volume of the subrange containing x is the volume of the corresponding parallelepiped X_j. In the coordinate system $y_1, \ldots, y_n$ formed by unit eigenvectors of the matrix c_{ab}, this matrix has a diagonal form $c'_{aa} = \lambda_a$ and $c'_{ab} = 0$ for $a \ne b$, where λ_a is the corresponding eigenvalue of the original matrix c_{ab}. In these coordinates, the condition $\Delta G_j(x) \le v$ takes the form $\sum\limits_{a} \lambda_a \cdot (\Delta y_a)^2 \le v$. Thus, each axis has half-length $\sqrt{\dfrac{v}{\lambda_a}}$ and, correspondingly, length $2\sqrt{\dfrac{v}{\lambda_a}}$. The volume of the box X_j is equal to the product of these lengths, i.e., to

$$2^n \cdot v^{n/2} \cdot \sqrt{\frac{1}{\prod\limits_{a} \lambda_a}}. \tag{15}$$

The product of all the eigenvalues of a matrix is equal to its determinant, so $\prod\limits_{a} \lambda_a = \det(c_{ab})$. Thus, the volume of each subrange is equal to

$$v = 2^n \cdot v^{n/2} \cdot \frac{1}{\sqrt{\det(c_{ab})}}. \tag{16}$$

So, in a unit volume close to the point x, we have $1/v$ such sub-ranges, i.e.,

$$\sqrt{\det(c_{ab}(x))} \cdot 2^{-n} \cdot v^{-n/2} \tag{17}$$

subranges. The overall number of subranges can be obtained if we add these numbers over all unit-volume parts of the range X, i.e., if we consider the integral

$$\int \sqrt{\det(c_{ab}(x))} \cdot 2^{-n} \cdot v^{-n/2} \, dx. \tag{18}$$

The number of subranges should be equal to N, so we conclude that

$$\int \sqrt{\det(c_{ab}(x))} \cdot 2^{-n} \cdot v^{-n/2} \, dx = N, \tag{19}$$

and thus, that

$$v^{n/2} = \frac{2^{-n}}{N} \cdot \int \sqrt{\det(c_{ab}(x))} \, dx, \tag{20}$$

so

$$v = \frac{1}{4 \cdot N^{2/n}} \cdot \left(\int \sqrt{\det(c_{ab}(x))} \, dx \right)^{2/n}. \tag{21}$$

Thus, we arrive at the following optimal division into subranges.

Solution: optimal division into subranges. Suppose that we can have N subranges. Then, we compute the value v by using the formula (21), where the matrix $c_{ab}(x)$ is determined by the formula (13). This value v is the largest possible difference between the optimal and suboptimal values of the objective function $G(x, u)$.

The corresponding subranges have the following form. Around each point x, we find the unit eigenvectors and eigenvalues λ_a of the matrix $c_{ab}(x)$. In the local coordinate system $y_1, \ldots, y_n$ formed by the unit eigenvectors, the subrange X_j is the following box:

$$\left[y_1^{(j)} - \sqrt{\frac{v}{\lambda_1}}, y_1^{(j)} + \sqrt{\frac{v}{\lambda_1}} \right] \times \ldots \times \left[y_n^{(j)} - \sqrt{\frac{v}{\lambda_n}}, y_n^{(j)} + \sqrt{\frac{v}{\lambda_n}} \right]. \tag{22}$$

4. Case When We Know the Probabilities of Different Situations

Description of the case. Suppose that we also know the relative frequency of different situations, i.e., we know the probability density function $\rho(x)$ describing how frequently we will encounter different situations.

Analysis of the problem. In this case, as one can show, locally, we have a similar division. The difference is that instead of the same value v_j for all the subranges, we may have different values $v(x)$ for different subranges: to decrease the average measure of difference $\Delta G_j(x)$, it makes sense to make it larger for scarcely populated subranges and smaller for densely populated subranges.

Once we know $v(x)$ for each x, we can determine the corresponding division into subranges. So, the main remaining problem is finding the optimal function $v(x)$. The main constraint is the overall number N of subranges, which, similar to formula (19), has the form

$$2^{-n} \cdot \int \sqrt{\det(c_{ab}(x))} \cdot (v(x))^{-n/2} \, dx = N. \tag{23}$$

Under this constraint, we want to minimize the average difference $\Delta G(x)$. For each subrange, the average difference is proportional to $v(x)$, so minimizing the average difference is equivalent to minimizing the average values of $v(x)$:

$$\int \rho(x) \cdot v(x) \, dx \to \min. \tag{24}$$

By using the Lagrange multiplier method, we can reduce this constraint optimization problem to the unconstrained problem of minimizing the functional

$$\int \rho(x) \cdot v(x) \, dx + \lambda \cdot \left(2^{-n} \cdot \int \sqrt{\det(c_{ab}(x))} \cdot v(x)^{-n/2} \, dx - N \right), \tag{25}$$

where λ is the Lagrange multiplier. Differentiating this expression with respect to $v(x)$ and equating the derivative to 0, we conclude that

$$\rho(x) - \lambda \cdot 2^{-n} \cdot \frac{n}{2} \cdot \sqrt{\det(c_{ab}(x))} \cdot (v(x))^{-n/2-1} = 0, \tag{26}$$

i.e., that:

$$(v(x))^{-n/2-1} = C \cdot \frac{\rho(x)}{\sqrt{\det(c_{ab}(x))}}, \tag{27}$$

where we denoted

$$C \overset{\text{def}}{=} \frac{1}{\lambda \cdot 2^{-n} \cdot \dfrac{n}{2}}. \tag{27}$$

Thus,

$$(v(x))^{-n/2} = \left((v(x))^{-n/2-1}\right)^{n/(n+2)} =$$

$$C^{n/(n+2)} \cdot \frac{(\rho(x))^{n/(n+2)}}{\left(\sqrt{\det(c_{ab}(x))}\right)^{n/(n+2)}}. \tag{28}$$

Substituting this expression for $(v(x))^{-n/2}$ into the formula (23), we conclude that

$$2^{-n} \cdot C^{n/(n+2)} \cdot \int (\rho(x))^{n/(n+2)} \cdot (\det(c_{ab}(x)))^{1/(n+2)} \, dx = N, \tag{29}$$

hence

$$C^{n/(n+2)} = \frac{2^n \cdot N}{\int (\rho(x))^{n/(n+2)} \cdot (\det(c_{ab}(x)))^{1/(n+2)} \, dx}, \tag{30}$$

and

$$C = \frac{2^{n+2} \cdot N^{1+2/n}}{\left(\int (\rho(x))^{n/(n+2)} \cdot (\det(c_{ab}(x)))^{1/(n+2)} \, dx\right)^{1+2/n}}. \tag{31}$$

From (27), we can then conclude that

$$v(x) = \frac{(\det(c_{ab}(x)))^{1/(n+2)}}{C^{2/(n+2)} \cdot (\rho(x))^{2/(n+2)}}. \tag{32}$$

So, we arrive at the following solution.

Solution: optimal division into subranges. First, we compute the auxiliary value C by using the formula (31). Then, the corresponding subranges have the following form. Around each point x, we find the unit eigenvectors and eigenvalues λ_a of the matrix $c_{ab}(x)$. In the local coordinate system $y_1, \ldots, y_n$ formed by the unit eigenvectors, the subrange X_j is the following box:

$$\left[y_1^{(j)} - \sqrt{\frac{v(x)}{\lambda_1}}, y_1^{(j)} + \sqrt{\frac{v(x)}{\lambda_1}}\right] \times \ldots \times \left[y_n^{(j)} - \sqrt{\frac{v(x)}{\lambda_n}}, y_n^{(j)} + \sqrt{\frac{v(x)}{\lambda_n}}\right], \tag{33}$$

where $v(x)$ is determined by the formula (32).

Acknowledgments

This work was supported in part by the National Science Foundation grants 1623190 (A Model of Change for Preparing a New Generation for Professional Practice in Computer Science) and HRD-1242122 (Cyber-ShARE Center of Excellence).

References

Belohlavek, R., Dauben, J. W. and Klir, G. J. (2017). Fuzzy Logic and Mathematics: A Historical Perspective, Oxford University Press, New York.

Kahneman, D. (2013). Thinking, Fast and Slow, Farrar, Straus, and Giroux, New York.

Klir, G. and Yuan, B. (1995). Fuzzy Sets and Fuzzy Logic, Prentice Hall, Upper Saddle River, New Jersey.

Lorkowski, J. and Kreinovich, V. (2018). Bounded Rationality in Decision Making Under Uncertainty: Towards Optimal Granularity, Springer, Cham, Switzerland.

Mendel, J. M. (2017). Uncertain Rule-Based Fuzzy Systems: Introduction and New Directions, Springer, Cham, Switzerland.

Nguyen, H. T., Walker, C. L. and Walker, E. A. (2019). A First Course in Fuzzy Logic, Chapman and Hall/CRC, Boca Raton, Florida.

Novák, V., Perfilieva, I. and Močkoř, J. (1999). Mathematical Principles of Fuzzy Logic, Kluwer, Boston, Dordrecht.

Zadeh, L. A. (1965). Fuzzy sets. Information and Control, 8: 338–353.

Chapter 5

Embodied 3D Neural Tissue Cultures for Cognitive Research

Nicolas Rouleau

1. Introduction

Functions of the brain have historically been classified as either behavioral or cognitive with the implicit assumption that one set of capacities can be isolated from the other without compromising their most essential and discriminant features. Consequently, the field of psychology has undergone significant paradigm shifts over the past century, oscillating between frameworks centered on mental processes or, alternatively, observable motor responses. Since, subjective experiences associated with cognitive processes remain opaque to direct, empirical measurement, contemporary investigators continue to rely on correlative factors (Koch et al., 2016; Llinás et al., 1998; Meador et al., 2002). Indeed, cognition is either inferred on the basis of highly correlated behavioral patterns such as verbal self-report or by observable brain activations with neuroimaging tools (i.e., reverse inference) such as functional magnetic resonance imaging (fMRI) or electroencephalography (EEG) (Poldrack, 2006). It is uncontroversial to suggest that activations of particular brain structures are reliably associated with specific functional outputs and experiences. However, the brain's inordinately complex microcircuitry represents an enduring bottleneck toward a

Department of Biomedical Engineering, Tufts University, Science and Technology Center, 4 Colby Street, Medford, Massachusetts, USA.
Department of Psychology, Algoma University, Shingwauk Hall, 1520 Queen St. E., Sault Ste. Marie, Ontario, Canada.

comprehensive understanding of how defined neural ensembles give rise to higher-order cognitive processes at a mechanistic level. With the goal of understanding the mental functions of naturally occurring and self-organized brains in living organisms, it may be necessary to first re-direct our attention to artificial equivalents as tools for discovery.

Neuroscientists are consistently generating enormous volumes of information on cell signaling, gene expression, network dynamics, and other detailed features of brains that are undoubtedly relevant to cognition but reliably beg the "big picture" questions to which humans *actually* want answers: What are the necessary and sufficient neural conditions from which experience emerges? Are constructs like "love" and "justice" reducible to defined network properties? How are complex and often self-contradictory rules encoded to define personal morality? How do neural tissues distinguish between veridical and imagined perceptions (if they do at all)? How do brains represent "meaning" in their microcircuitry? Our failure to grapple with the substance of these questions likely stems from an inability to shed old habits. One of the main limitations of our current approach—which has admittedly served the neuroscientific community well as a simple fact of history—is an assumed intrinsic importance associated with specific neural structures, biomolecules, and signal patterns as they relate to mental functions. In general, the biological sciences tend to preoccupy themselves with *substrate-dependent* properties of organisms rather than attempting to find generalizable principles, rules, equations, and other fundamental operations that independently explain the phenomenon with the assumption that parts of the whole can be substituted or replaced without compromising outcomes. Until we replace our reverence for "natural", ontogenetic morphology as a standard model for cognitive research with an open-minded exploration of artificial analogues, investigative potential in cognitive research will be limited.

In recent years, several authors have suggested that cognition may be a ubiquitous property of living systems that is fundamentally *substrate-independent* (Bostrom, 2003; Moon and Pae, 2019). Indeed, operant mechanisms underlying cognition may be fundamental to simple chemical reactions at the root of life itself. Single-celled organisms can be conditioned (Hennessey et al., 1979), navigate complex environments (Lyon, 2015; Murugan et al., 2021; Zhu et al., 2013), solve mazes (Nakagaki et al., 2000), and display decision-making behaviors that are operationally indistinguishable from those of mammals including humans (Lyon, 2020; Lyon et al., 2021; Zhu et al., 2013). From bacteria to insects and plants, there is considerable evidence that information processing—a fundament of cognition—is ubiquitous and expressed across the full spectrum of living systems (Lyon, 2020; Lyon et al., 2021). The term "minimal cognition" is becoming increasingly popular as a conceptual placeholder for what is an emerging science of thought and experience at the most rudimentary level (Lyon, 2020). With the goal of identifying universal operations that transcend species-specific tissue substrates, the field of minimal cognition has been extended to the study of dissociated neural cultures and tissue preparations *in vitro*. These systems are simple, flexible, and can be systematically manipulated to potentially determine the causes rather than the correlates of cognition. However, because—like the brain—haphazard neural activity "in a dish" is effectively meaningless without bodies with which to "decode the message", new approaches will be needed to fuel discovery. Fortunately, advances in neural-computer interfaces and robotics are bridging these gaps and clearing a path toward a more sophisticated understanding of cognition as a generalized functional property of many systems.

2. What are Neural-computer Interfaces?

Neural-computer interfaces (NCIs)—also called brain-computer interfaces or BCIs—are integrated systems that pair neural cells and tissues with computers, robots, machines, and other artificial outputs that record activity or couple signals to auxiliary resources to do work (Grosse-Wentrup et al., 2011; Halder et al., 2011; Soekadar et al., 2015) (Fig. 1). Whereas outputs of the central nervous system typically activate other CNS cells (thought) or muscles by way of the peripheral nervous system (behavior), NCIs replace these terminal effectors with artificial sensors and actuators.

The basic principle of the NCI was first demonstrated in the 1920s by Hans Berger when he invented the EEG (Millett, 2001)— a device that converted voltages over the surface of the scalp into mechanical oscillations of a needle (neural-machine interface). The needle, which exuded a stream of ink, traced the neural oscillations on paper as a temporal record that could be quantified to infer cognitive states or functional pathologies like seizures. Since Berger's time, the EEG has been fully digitized and is still used today as a common instrument in cognitive psychology and neuroscience research as well as in clinical practice.

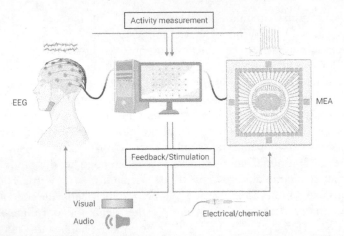

Figure 1. Neural-computer interfaces (NCIs). Electrophysiological (EEG, MEA) or optical (not pictured) signals associated with neural activations are measured and analyzed to trigger feedback. Stimuli can feedback visual, audio, electrical, or chemical signals to generate a closed loop to facilitate learning. This figure was created with BioRender.com.

Today, NCIs are becoming increasingly popular and available as therapeutic devices and personal tools. Neurofeedback training, which pairs brain activity with computers that deliver threshold-triggered sensory feedback (Fig. 1) have allowed clinical patients to consciously identify and learn to suppress their own seizures (Sterman and Egner, 2006; Walker and Kozlowsky, 2005), increase their concentration (Chung et al., 2008; Hammond, 2011), and alleviate anxiety (Hammond, 2005). Commercial NCIs are currently available as sleep aids and recreational gaming controllers. Industry leaders have even suggested that, within the decade, minimally invasive brain implants could provide humans with the ability to control their handheld and other electronic devices with thought alone (Pisarchik et al., 2019)—reifying technologies that were previously relegated to the status of science fiction. Implants signaling wirelessly to intermediate devices or other implants with on board processors could soon increase the efficiency and proficiency of human-to-human communication. Circumventing historical limitations to human communication such as geographical distance, these devices would effectively tighten human social networks by contracting spatial and temporal constraints of social behavior.

Beyond commercial applications, NCIs are providing researchers with flexible, modular, and iterative solutions to study artificial intelligence (AI) and cognitive processes such as learning and memory. Indeed, optogenetic and other modern brain-stimulation technologies use light or voltage to trigger neuronal activity, thus modifying an organism's cognitive and behavioral patterns (Kravitz and Kreitzer, 2011; Zhang et al., 2010). These changes can be recorded with arrays of implanted electrodes within the stimulated region to detect functional neural correlates and elucidate brain-based mechanisms of functions at the organismal level. Paralleling these developments, biomedical engineers have made significant progress with *in vitro* NCIs. Coupling neural tissue cultures derived from the dissociated cells of brains from formerly living organisms with "artificial bodies" as simulated or real-world machines has revealed that *in vitro* NCIs make their own meaning by interacting with an external world that can be encoded as network-bound representations (DeMarse and Dockendorf, 2005). These techniques are allowing researchers to partially overcome the overwhelming swell of reductionist findings in the ever-expanding neuroscience literature with top-down models of how cognitive systems self-organize and change as a function of information feedback loops (Potter, 2007).

3. Embodied Cultured Networks

It was previously assumed that bodies were hostage to mental action—they did as they were told and provided mindless sensory feedback to improve the one true cognitive control center: the brain. However, contemporary theorists are now exploring the possibility that cognition is fundamentally embodied (Wilson, 2002). This radical idea suggests not only that the body and extended environment of the organism influence cognition or that cognition reciprocates, but also that the full, interconnected, and distributed chain of brain-body interfaces constitute a collection of cognitive resources (Anderson, 2003; Wilson and Golonka, 2013). Consistent with the idea of minimal cognition, embodied cognition implies many decentralized cognitive agents—not just brains— contribute to behavior (Wilson and Golonka, 2013). As previously discussed, subjective experiences can only be inferred if the brain is functionally coupled to a body that can decode and translate mental representations into objective, executable response patterns. In other words, cognition is not quite reducible to the flow of information within a system. Rather, understanding a cognitive process requires a thorough appraisal of the task to be solved, the resources at the system's disposal, and its capacity to assemble and coordinate said resources toward solving the task (Wilson and Golonka, 2013).

Applied to cultured neural networks, embodiment begins with the measurement of activity from both individual and groups of cells. In principle, any feature of neural activity—from the influx of ions, to the action potential, and the downstream release of neurotransmitters— can serve as the signal source. Consequently, both optical and electrophysiological detectors are reported in the literature. However, multielectrode arrays (MEAs) represent the most common interfaces, reinforced by over 40 years of technical development (Pine, 1980; Kim et al., 2018). Local field potentials (LFPs), recorded over dozens of distributed electrodes associated with the generation of local action potentials, are measured as punctate voltage fluctuations that vary by amplitude and frequency over time (Herreras, 2016). Using analog-to-digital converters, the spatial array of signals is digitized and sent to a computer that processes the information. Clusters of MEA sensors and their overlayed cells can be pre-defined as "motor" or "sensory" regions to output signals or receive stimulation as needed (Cozzi et al., 2005). This procedure partitions the neural network into functionalized regions rather than treating the culture as a homogenous mass. Whereas naturally occurring brains are equipped with morphogenetically conserved, functionalized brain regions, subsections of cultured neural networks must be defined by the experimenter based upon the input-output design of sensors and actuators.

It should be noted that brains without bodies or NCIs without sensory feedback to the neural preparation, are functionally abnormal. Interestingly, when disembodied neural networks are metabolically sustained without sensory feedback, they produce transient, significantly dilated action potentials (~ 100 ms/spike) that recur every few seconds—an abnormal signaling pattern not observed *in vivo* (Potter et al., 2004). However, delivering multi-site sensory feedback reverses the dysfunction and restores electrophysiological normalcy (Madhavan et al., 2003; Potter et al., 2004). It is likely that because neural systems evolved to control bodies that provided steady streams of sensory feedback, they developed dependencies that have persisted phylogenetically (Clark, 1998; Potter et al., 2004). Disembodied neural tissues are therefore intrinsically aberrant systems that are cutoff from essential resources that allow them to perform higher-order functions. Indeed, the presence of bodily feedback as background noise is essential to stabilizing networks under pulsed stimulation conditions (tetanization) which otherwise devolve into states of hypersynchronous, seizure-like electrical paroxysms or "barrages" (Chao et al., 2005). Therefore, "re-embodying" dissociated neural cultures provides a necessary element that is fundamentally missing from the network itself—a testament to the relevance of embodied cognition theory.

When engineering an embodied neural network (Fig. 2), "closing the loop" (Potter et al., 2006) between outgoing (motor) and incoming (sensory) signals permits an experience-dependent wiring of the network (Weiler et al., 1995), also known as *learning*. Indeed, increased performance indicative of learning associated with an

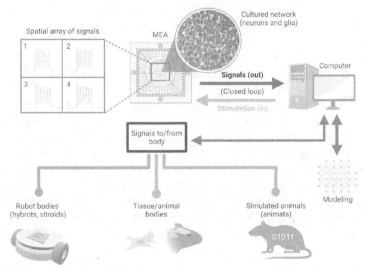

Figure 2. Embodied cultured networks control robots, tissues, and simulations. Spatial arrays of signals from cultured neural networks (neurons and glia) are detected by multi-electrode arrays (MEAs) where outputs to a computer either trigger events or are processed by modeling software or artificial neural networks. Computers then trigger motor actuators associated with virtual or real-world bodies including mobile robots, animal bodies (biological tissues with injections of current), or simulated animal bodies (animats). Sensory feedback from bodies ultimately returns to the cultured network to provide the necessary information to learn and adapt. Plasticity within the embodied neural network, facilitated by electric pulses that track sensory feedback, changes synaptic strengths such that the microcircuitry becomes tuned to the task, optimizing performance. The closed sensory-motor loop is the key element where outputs are contingent upon a history of inputs. This figure was created with BioRender.com.

in vitro cortical network coupled to an MEA with real-time feedback (i.e., an NCI) was fully inhibited by cutting off sensory feedback (Bakkum et al., 2008). If, as theoretical works suggest, higher-order cognitive processes such as intelligence can be fundamentally operationalized as the maximization of reward (Silver et al., 2021), learning is the ultimate cognitive linchpin because capturing rewards will always depend on an acquired understanding of the environment. But how does a cultured neural network know what the sensory data is representing? How does the network distinguish itself from the environment? What is it like to be a brain in a dish? Of course, there's no reason to suspect that any organism requires an understanding of the world as it *actually is* to compete and thrives within it (Hoffman et al., 2015; Prakash et al., 2020). In fact, the evidence suggests that a true or veridical perception of the environment is less desirable than practical fitness outcomes regardless of the resultant perceptual distortion (Hoffman et al., 2015; Prakash et al., 2020). In other words, usefulness out competes truth. Therefore, encoding and decoding neural activity in NCIs does not need to perfectly track objective referents in the environment. Rather, key features inherent to network activity and environment can be used to guide sensory-motor feedback loops toward pre-defined goals.

Decoding the activity of cultured neural networks involves the identification of signal features that can be exploited to define motor outputs (Tessadori et al., 2013). Individual voltage spikes, combinations of spikes (doublets, triplets), inter-spike intervals, bursts, and other patterns have been reported in the literatures (Tessadori et al., 2013); however, the evidence suggests that these discrete signal features do not necessarily contain sufficient information to drive improved performance (Tessadori et al., 2013). Indeed, the whole signal may be greater than the sum of its parts. In one study, an *in toto* approach was used to extract the maximum information content from the unpartitioned signals of neural cultures which were transmitted first to an Artificial Neural Network (ANN) and then to a robot to perform a task (Pizzi et al., 2009). The ANN's precise computational strategies were obscured like a cognitive "black box"; however, it was clear that network activity could be decoded with high accuracy to perform robotic tasks. Whatever signal feature or combination of features enhanced performance was, in fact, present in the activity of the network even if it was recondite. Interestingly, the effect of the ANN was only apparent once the cells within the cultured neural network were trained using sensory feedback. Which is to say, the information-content of the embodied culture was only discernable by the ANN once it had been functionally mapped by learning processes driven by the NCI's sensory feedback (Pizzi et al., 2009).

One interpretation of this remarkable discovery is that neural ensembles make their own decodable meaning by learning processes that encode feedback as a representation of the extended environment. However, functionalized regions can be intentionally engineered into the network by defining "sensory" and motor" regions *a priori* as a function of space. For example, left and right halves or center-surround organizations of the cultured neural networks can be programmed to pre-define sensory and motor regions (Tessadori et al., 2012). In principle, any geometry may provide the means of spatial encoding but there are likely quantitative limitations on the bases of spatial overlap and raw cell numbers. Nevertheless, there are two typical strategies that are currently used to encode information within embodied neural cultures as NCIs that display electrical activity—the first of which involves Spike-Timing Dependent Plasticity (STDP).

The theoretical foundation for STDP is rooted in Hebbian learning (Lobov et al., 2021; Roberts and Leen, 2010) where, as Donald Hebb described himself, "neurons that fire together, wire together". Stated otherwise, when activity between pre- and post-synaptic neurons becomes highly correlated or synchronous, the connections between them strengthens markedly. With STDP, it is possible to entrain synchronous activity within a neural circuit (Nowotny et al., 2003). Practically speaking, this involves an application of high-frequency stimuli—usually trains of pulsed voltage—to the network with frequencies greater or equal to 20 Hz. In simulations of Spiking Neural Networks (SNNs), STDP promotes the encoding of localized "negative" or "dangerous" stimuli in the environment to be avoided (Lobov ct al., 2021). In experiments with cortical NCIs involving real-world mobile robot bodies, 20 Hz pulses were applied over a 2 second period to the side of the cortical network that was pre-defined as "sensory" and associated with the collision-side sensors of the mobile robot body (Novellino, 2007). The authors hypothesized that by strengthening synaptic connections among poorly correlated regions of the network, compensation as avoidance would be possible with training. Indeed, they observed short term plasticity associated with the STDP-type protocol with increased high frequency activity within the network that modulated motor outputs for increased performance (Jimbo et al., 1998). Similar results were reported using a similar technique with hippocampal rather than cortical cultures (Tessadori et al., 2012).

The second technique, which is termed anti-STDP (anti-Hebbian), involves the weakening of presynaptic inputs to post-synaptic targets and is dependent upon strict timing constraints (Roberts and Leen, 2010). Rather than promoting "avoidance" anti-STDP promoted the localization of "positive" or "desirable" stimuli in SNNs (Lobov et al., 2021). One real-world application of STDP and anti-STDP principles involves potentiating increased network response when the second of two injected stimuli are presented less than 30 ms (> 33.3 Hz) apart (STDP) or depressed (anti-STDP) when they were 100–500 ms apart (2 Hz–10 Hz) (Shkolnik, 2003; Potter et al., 2004). In the context of NCIs, the goal of both STDP and anti-STDP is to change the network's connective properties to reflect some feature of the environment informed by sensory feedback. Together, these techniques can be used to program NCIs with many possible applications.

4. Cybernetic Organisms

Embodied cultures as NCIs are hybrid organisms composed of biological and artificial or electric components, which satisfy the classic definition of a cybernetic organism or "cyborg". Steve M. Potter's group's hybrid robots or "hybrots", developed at the George Institute of Technology, are among the most well-known cyborgs that have been developed. Among them, the Koala 6-wheeled rover (K-Team) is the most iconic. It can approach and track a second, randomly operated rover based on sensory feedback coded as short or long interspike intervals which maps the distance between the two robots (Potter et al., 2004; Shkolnik, 2003). Another notable hybrot is the MEART, which is a "semi-living artist" composed of a robotic arm controlled by 50,000 cultured neurons on an MEA that creates drawings with imaging feedback from a CCD camera (Bakkum et al., 2007; Potter et al., 2004). Both systems are based upon the same NCI principles involving cultured neural network monolayers on MEAs and can be trained to increase performance. As will now be discussed, a shift within the field is currently underway that will increase the complexity of NCIs with the potential to generate high-functioning cyborgs for cognitive research.

5. The 2D-3D Ttransition: Bioengineering Mini Brain Hybrots

Embodied NCIs have classically involved two-dimensional (2D) neural monolayers of dissociated cells from formerly living brains; however, in recent years there has been some effort to transition from 2D cell culture to three-dimensional (3D) neural tissue cultures. In particular, the replacement of monolayers with organotypic slice cultures has provided investigators with new potential directions for the construction of complex NCIs. For example, hybrots have been developed with thick (500 μm) neural slice preparations using custom multiphoton imaging systems (Rambani, 2007) and MEAs (Killian et al., 2016). Indeed, the development of convection-based superfusion techniques has enabled the long-term (e.g., weeks, months, years) sustainability of slice cultures that previously suffered from necrosis associated with poor nutrient and oxygen diffusion (Killian et al., 2016). Organotypic slice cultures contain defined circuits that are intrinsic to ontogeny, which means they display conserved microcircuitry with well-defined functions.

The transition from monolayers to slice preparations is an exciting step forward for the field; however, slice cultures suffer from significant limitations that cannot simply be mitigated by the application of ever-more sophisticated culturing techniques. First and foremost, organotypic slice cultures are intrinsically damaged by dint of having been severed from their former efferent and afferent connections in the brain (Rouleau et al., 2021). This means that all connections that are directionally orthogonal to the injury plane are effectively lost and are not sufficiently plastic to re-integrate with surrounding tissues. Therefore, large sections of the tissues serve little or no function. Second, slices are derived from brains with pre-defined, functionalized brain regions. For example, a slice may contain motor and sensory cortices that were previously connected to descending or ascending tract systems within an animal body. In other words, neural networks within slice preparations are not experience-naïve in the same way dissociated cells are or are made to be– rather, they are fundamentally biased by a history of inputs and outputs that define the network's synaptic connectome. Therefore, if the transition from 2D-to-3D tissues is to continue within the field, investigators must move away from slice preparations and toward more flexible systems that can be controlled and defined.

Bioengineered 3D neural tissues are becoming increasingly accessible as tools for the cognitive sciences (Rouleau et al., 2021; Rouleau et al., 2021). Many "mini brain" models have been generated in recent years using bioprinting, spheroids, organoids, scaffold-based tissues, organ-on-a-chip configurations, and other 3D constructs (Gu, 2015; Rouleau et al., 2021). In general, 3D neural tissues are more physiologically relevant as they provide greater surface area for cell-cell interactions, increased cell density, and more realistic macro-scale signaling dynamics. Additionally, the composition of 3D neural tissues can be systematically controlled including the makeup of the extracellular matrix as well as the integration of cooperative cell types including vasculature for realistic perfusion of gases and small molecules. As measured by standard MEA technology, cortical organoids have displayed human-EEG-like oscillatory profiles which suggests stable network properties with synchronous activity can be generated in 3D mini brains (Trujillo et al., 2019). Paralleling advances in tissue culture, there has been a surge in novel 3D MEA designs in the past 5 years with approaches that either incorporate embedded vertical arrays, protruding rods, sandwich-type configurations, or conformal layering of electrodes (Choi et al., 2021). It is predicted that complex 3D network features will soon become detectable, inspiring the next generation of 3D NCIs with applications in robotics and AI.

Applied to investigations of embodied cognition, 3D NCIs are likely to provide increased physiological relevance and additional dimensions of complexity, control, tunability, and programmability. Indeed, as 3D bioengineered neural tissues become coupled with cybernetic bodies, the resulting cyborg hybrids could serve as the next generation of laboratory model systems with adapted behavioral paradigms for mini brain cognition experiments (Fig. 3). Some leading investigators have already designed and fabricated 3D neuromuscular organoids with and without spinal cord-like intermediates that generate motor responses as a function of mini brain network activity (Andersen et al., 2020; Bakooshli et al., 2019; Martins et al., 2020). Therefore, top-down motor control is already possible with 3D *in vitro* systems. Similarly, neural organoids with optic vesicles have been created that are sensitive to light (Gabriel et al., 2021); thus, sensory feedback is also possible. Now that the major pathway associated with voluntary movement—the corticospinal tract – has been recapitulated *in vitro* as a self-organizing

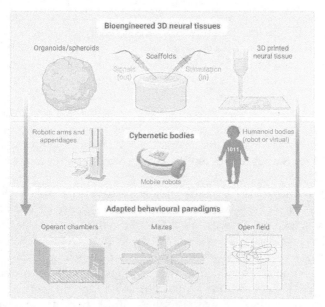

Figure 3. Embodying bioengineered 3D neural tissues for cognitive research. Organoids as well as scaffold-based and bioprinted 3D tissues equipped with cybernetic bodies will soon require adapted behavioral paradigms with applications in cognitive research. 3D NCIs and hybrots are anticipated as the next generation of model systems with which to assess mechanisms of cognition in the laboratory. This figure was created with BioRender.com.

"assembloid" and investigators can already close the sensory-motor loop with purely biological substrates, there is little doubt that next-generation 3D NCIs will match and even surpass these developments within years.

6. Applications for Minimal Cognition Research

In 2021, Rouleau et al., published a review entitled *Toward Studying Cognition in a Dish* that discussed the future of embodied cultures in cognitive research. The authors argued that recapitulating higher-order cognitive functions such as decision-making, abstract reasoning, and even consciousness *in vitro* would provide investigators with the means of reverse-engineering the "black box" of the mind. Because, unlike animal models and dissociated monolayer cultures, bioengineered neural tissues can be designed with limitless iterations of unique cytoarchitectural organization and cell type composition, there is significant potential to identify fundamental substrate principles that are necessary or sufficient for cognitive function. Indeed, it was predicted that by adding and subtracting tissue elements, investigators would soon be able to elucidate and instruct a new causal model of neurocognition that transcends current standards of correlative functional anatomy. Even a capacity for consciousness may be identifiable by constructing embodied hybrid systems that are both experimentally flexible and subject to stimulus-response-based query. If consciousness is fundamentally a product of brain activity, the possibility of subjective experience in 3D neural tissues, that already display many neural correlates of consciousness including known EEG rhythms indicative of awareness, must be considered with some level of sincerity. However, the authors argued that before any such investigation could occur, it was first necessary to design appropriate interfaces with which to decode neural activity and translate it into motor responses with which to infer mental processes.

The first generation of adapted behavioral assays for 3D NCIs involving bioengineered neural tissues will likely center on learning and memory as well as comparatively simple displays of minimal cognition in hybrid systems. Currently, there are only a few examples of *in vitro* cognitive assays using 3D bioengineered neural tissues. One recent paper demonstrated that stimulation-induced habituation, which is a form of non-associative learning, was displayed by 3D cortical networks (Rouleau et al., 2021). The networks also displayed spontaneous recovery, which indicates that the functional plasticity was short-term. However, as electronic interfaces become more accessible, many more demonstrations are to be expected. Subsequent developments of minimal cognitive research *in vitro* will likely borrow from the behaviorist tradition and involve highly operationalized procedures that involve decision-making-based outputs. Indeed, adapting classic operant chambers for *in vitro* work in general but 3D NCIs in particular would provide investigators with standardized means of assessing embodied cognitive task performance with quantifiable outcomes (Fig. 3). Since the tissues are highly modular, it may be possible to derive powerful laws, principles, and predictive patterns that underlie intelligent systems and their decision-making operations. Of course, one distant but desirable goal would be to identify a subset of key elements that can explain subjective experience or consciousness in minimally cognitive systems. With embodied 3D cultures, it may be possible to know what simple tissues *can know* by designing Turing tests for cyborgs. Interpreting results from any such test will require extreme caution; however, the prospect of building brains to think, behave, and experience is one that should arouse excitement in any cognitive scientist.

7. Ethical and Legal Considerations

A sincere appreciation for the implications of potentially conscious embodied neural tissue preparations—bioengineered or otherwise—demands some consideration for ethical and legal issues. Krausová and Hazan (2017) broke ground in this area, discussing issues of liability in cases of hybrots causing damage to people or property, as well as the social and ethical impacts of general conflicts with cyborgs. The authors suggest that responsibility, which hinges on definitions of control and a capacity for moral decision-making, is a challenging concept when applied to AI and other non-human agents. In the case of hybrots, their creators or those who influence or determine their actions may be held responsible, but this will likely also depend upon the design of the system and its level of programmed autonomy. The freedom to overwrite rules, which is a feature of any cultured neural network capable of synaptic plasticity, will further complicate ethical and legal considerations. If a cyborg is programmed to write its own code or modify its neural networks, is the programmer ultimately responsible for the downstream consequences of the initial decision to forfeit control? Anticipating these technologies will eventually mature sufficiently to display human-level capabilities or surpass them, Krausová and Hazan (2017) suggest that it is reasonable to reserve the status of "electronic person" or "hybrid person" for hybrots and similar systems. Whether the locus of control is of a biological origin or not may be a critical determinant of personhood though a sufficiently advanced AI may equally qualify. Of course, the ability to harm hybrots and cyborgs or cause them to suffer will depend upon the presence of certain types of receptors (e.g., nociceptors), the ability to perceive or anticipate pain, a capacity for emotionality, a capacity for consciousness and self-awareness, memory of traumatic events and other relevant factors. Designing cyborgs therefore represents an intrinsic ethical dilemma since elements can be added or omitted by the engineer *a priori* with some knowledge of the implications of each modification.

References

Andersen, J., Revah, O., Miura, Y., Thom, N., Amin, N. D. et al. (2020). Generation of functional human 3D cortico-motor assembloids. Cell, 183(7): 1913–1929.

Anderson, M. L. (2003). Embodied cognition: A field guide. Artificial Intelligence, 149(1): 91–130.

Bakkum, D. J., Chao, Z. C. and Potter, S. M. (2008). Spatio-temporal electrical stimuli shape behavior of an embodied cortical network in a goal-directed learning task. Journal of Neural Engineering, 5(3): 310.

Bakooshli, M. A., Lippmann, E. S., Mulcahy, B., Iyer, N., Nguyen, C. T. et al. (2019). A 3D culture model of innervated human skeletal muscle enables studies of the adult neuromuscular junction. Elife, 8: e44530.

Bostrom, N. (2003). Are we living in a computer simulation? The Philosophical Quarterly, 53(211): 243–255.

Chao, Z. C., Bakkum, D. J., Wagenaar, D. A. and Potter, S. M. (2005). Effects of random external background stimulation on network synaptic stability after tetanization. Neuroinformatics, 3(3): 263–280.

Choi, J. S., Lee, H. J., Rajaraman, S. and Kim, D. H. (2021). Recent advances in three-dimensional microelectrode array technologies for *in vitro* and *in vivo* cardiac and neuronal interfaces. Biosensors and Bioelectronics, 112687.

Chung, S. Y., Seo, J. W., Kim, J. W. and Hwang, W. W. (2008). The effects of breath meditation with neurofeedback on memory and concentration of healthy adult volunteers. Journal of Oriental Neuropsychiatry, 19(2): 15–39.

Clark, A. (1998). Being There: Putting Brain, Body, and World Together Again. MIT press.

Cozzi, L., D'Angelo, P., Chiappalone, M., Ide, A. N., Novellino, A. et al. (2005). Coding and decoding of information in a bi-directional neural interface. Neurocomputing, 65: 783–792.

DeMarse, T. B. and Dockendorf, K. P. (2005, July). Adaptive flight control with living neuronal networks on microelectrode arrays.

In Proceedings 2005 IEEE International Joint Conference on Neural Networks, 2005. (Vol. 3, pp. 1548–1551). IEEE.

Gabriel, E., Albanna, W., Pasquini, G., Ramani, A., Josipovic, N. et al. (2021). Human brain organoids assemble functionally integrated bilateral optic vesicles. bioRxiv.

Grosse-Wentrup, M., Mattia, D. and Oweiss, K. (2011). Using brain–computer interfaces to induce neural plasticity and restore function. Journal of Neural Engineering, 8(2): 025004.

Gu, X. (2015). Progress and perspectives of neural tissue engineering. Frontiers of Medicine, 9(4): 401–411.

Halder, S., Agorastos, D., Veit, R., Hammer, E. M., S. Lee et al. (2011). Neural mechanisms of brain–computer interface control. Neuroimage, 55(4): 1779–1790.

Hammond, D. C. (2005). Neurofeedback with anxiety and affective disorders. Child and Adolescent Psychiatric Clinics, 14(1): 105–123.

Hammond, D. C. (2011). What is neurofeedback: An update. Journal of Neurotherapy, 15(4): 305–336.

Hennessey, T. M., Rucker, W. B. and McDiarmid, C. G. (1979). Classical conditioning in paramecia. Animal Learning & Behavior, 7(4): 417–423.

Herreras, O. (2016). Local field potentials: Myths and misunderstandings. Frontiers in Neural Circuits, 10: 101.

Hoffman, D. D., Singh, M. and Prakash, C. (2015). The interface theory of perception. Psychonomic Bulletin & Review, 22(6): 1480–1506.

Jimbo, Y., Robinson, H. P. and Kawana, A. (1998). Strengthening of synchronized activity by tetanic stimulation in cortical cultures: Application of planar electrode arrays. IEEE Transactions on Biomedical Engineering, 45(11): 1297–1304.

Kim, G. H., Kim, K., Lee, E., An, T., Choi, W. et al. (2018). Recent progress on microelectrodes in neural interfaces. Materials, 11(10): 1995.

Killian, N. J., Vernekar, V. N., Potter, S. M. and Vukasinovic, J. (2016). A device for long-term perfusion, imaging, and electrical interfacing of brain tissue *in vitro*. Frontiers in Neuroscience, 10: 135.

Koch, C., Massimini, M., Boly, M. and Tononi, G. (2016). Neural correlates of consciousness: Progress and problems. Nature Reviews Neuroscience, 17(5): 307–321.

Krausová, A. and Hazan, H. (2017). Robots with biological brains: Autonomy and liability of a semi-artificial life form. The Lawyer Quarterly, 7(3).

Kravitz, A. V. and Kreitzer, A. C. (2011). Optogenetic manipulation of neural circuitry *in vivo*. Current Opinion in Neurobiology, 21(3): 433–439.

Llinás, R., Ribary, U., Contreras, D. and Pedroarena, C. (1998). The neuronal basis for consciousness. Philosophical Transactions of the Royal Society of London. Series B: Biological Sciences, 353(1377): 1841–1849.

Lobov, S. A., Zharinov, A. I., Makarov, V. A. and Kazantsev, V. B. (2021). Spatial memory in a spiking neural network with robot embodiment. Sensors, 21(8): 2678.

Lyon, P. (2015). The cognitive cell: Bacterial behavior reconsidered. Frontiers in Microbiology, 6: 264.

Lyon, P. (2020). Of what is "minimal cognition" the half-baked version? Adaptive Behavior, 28(6): 407–424.

Lyon, P., Keijzer, F., Arendt, D. and Levin, M. (2021). Reframing cognition: Getting down to biological basics.

Madhavan, R., Wagenaar, D. A. and Potter, S. M. (2003). Multisite stimulation quiets bursts and enhances plasticity in cultured networks. In Society for Neuroscience Annual Meeting, 45–56.

Martins, J. M. F., Fischer, C., Urzi, A., Vidal, R., Kunz, S. et al. (2020). Self-organizing 3D human trunk neuromuscular organoids. Cell Stem Cell, 26(2): 172–186.

Meador, K. J., Ray, P. G., Echauz, J. R., Loring, D. W., Vachtsevanos, G. J. et al. (2002). Gamma coherence and conscious perception. Neurology, 59(6): 847–854.

Millett, D. (2001). Hans Berger: From psychic energy to the EEG. Perspectives in Biology and Medicine, 44(4): 522–542.

Moon, K. and Pae, H. (2019). Making sense of consciousness as integrated information: Evolution and issues of integrated information theory. Journal of Cognitive Science, 20(1): 1–52.

Murugan, N. J., Kaltman, D. H., Jin, P. H., Chien, M., Martinez, R. et al. (2021). Mechanosensation mediates long-range spatial decision-making in an aneural organism. Advanced Materials, 33(34): 2170263.

Nakagaki, T., Yamada, H. and Tóth, Á. (2000). Maze-solving by an amoeboid organism. Nature, 407(6803): 470–470.

Novellino, A., D'Angelo, P., Cozzi, L., Chiappalone, M., Sanguineti, V. et al. (2007). Connecting neurons to a mobile robot: An *in vitro* bidirectional neural interface. Computational Intelligence and Neuroscience, 2007.

Nowotny, T., Zhigulin, V. P., Selverston, A. I., Abarbanel, H. D., Rabinovich, M. I. et al. (2003). Enhancement of synchronization in a hybrid neural circuit by spike-timing dependent plasticity. Journal of Neuroscience, 23(30): 9776–9785.

Pine, J. (1980). Recording action potentials from cultured neurons with extracellular microcircuit electrodes. Journal of Neuroscience Methods, 2(1): 19–31.

Pisarchik, A. N., Maksimenko, V. A. and Hramov, A. E. (2019). From novel technology to novel applications: Comment on "An integrated brain-machine interface platform with thousands of channels" by Elon Musk and Neuralink. Journal of Medical Internet Research, 21(10): e16356.

Pizzi, R. M. R., Rossetti, D., Cino, G., Marino, D., Vescovi, A. L. et al. (2009). A cultured human neural network operates a robotic actuator. Biosystems, 95(2): 137–144.

Poldrack, R. A. (2006). Can cognitive processes be inferred from neuroimaging data? Trends in Cognitive Sciences, 10(2): 59–63.

Potter, S. M. (2007). What can AI get from neuroscience? In 50 Years of Artificial Intelligence (pp. 174–185). Springer, Berlin, Heidelberg.

Potter, A. M., DeMarse, T. B., Bakkum, D. J., Booth, M. C., Brumfield, J.R. et al. (2004). Hybrots: Hybrids of living neurons and robots for studying neural computation. Proceedings of Brain Inspired Cognitive Systems, 1–5.

Potter, S. M., Wagenaar, D. A. and DeMarse, T. B. (2006). Closing the loop: Stimulation feedback systems for embodied MEA cultures. In Advances in Network Electrophysiology (pp. 215–242). Springer, Boston, MA.

Prakash, C., Stephens, K. D., Hoffman, D. D., Singh, M. Fields, C. et al. (2020). Fitness beats truth in the evolution of perception. Acta Biotheoretica, 1–23.

Rambani, K. (2007). Thick brain slice cultures and a custom-fabricated multiphoton imaging system: progress towards development of a 3D hybrot model (Doctoral dissertation, Georgia Institute of Technology).

Roberts, P. D. and Leen, T. K. (2010). Anti-hebbian spike-timing-dependent plasticity and adaptive sensory processing. Frontiers in Computational Neuroscience, 4: 156.

Rouleau, N., Cairns, D. M., Rusk, W., Levin, M., Kaplan, D. L. et al. (2021). Learning and synaptic plasticity in 3D bioengineered neural tissues. Neuroscience Letters, 750: 135799.

Rouleau, N., Murugan, N. J. and Kaplan, D. L. (2021). Toward Studying Cognition in a Dish. Trends in Cognitive Sciences.

Shkolnik, A. C. (2003). Neurally Controlled Simulated Robot: Applying Cultured Neurons to Handle and Approach/Avoidance Task in Real Time, and a Framework for Studying Learning *in Vitro*. Masters Thesis, Emory University, Atlanta.

Silver, D., Singh, S., Precup, D. and Sutton, R. S. (2021). Reward is enough. Artificial Intelligence, 103535.

Soekadar, S. R., Witkowski, M., Vitiello, N. and Birbaumer, N. (2015). An EEG/EOG-based hybrid brain-neural computer interaction (BNCI) system to control an exoskeleton for the paralyzed hand. Biomedical Engineering/Biomedizinische Technik, 60(3): 199–205.

Sterman, M. B. and Egner, T. (2006). Foundation and practice of neurofeedback for the treatment of epilepsy. Applied Psychophysiology and Biofeedback, 31(1): 21.

Tessadori, J., Bisio, M., Martinoia, S. and Chiappalone, M. (2012). Modular neuronal assemblies embodied in a closed-loop environment: Toward future integration of brains and machines. Frontiers in Neural Circuits, 6: 99.

Tessadori, J., Venuta, D., Pasquale, V., Kumar, S. S., Chiappalone, M et al. (2013, July). Encoding of stimuli in embodied neuronal networks. In Conference on Biomimetic and Biohybrid Systems (pp. 274–286). Springer, Berlin, Heidelberg.

Trujillo, C. A., Gao, R., Negraes, P. D., Gu, J., Buchanan, J. et al. (2019). Complex oscillatory waves emerging from cortical organoids model early human brain network development. Cell Stem Cell, 25(4): 558–569.

Walker, J. E. and Kozlowski, G. P. (2005). Neurofeedback treatment of epilepsy. Child and Adolescent Psychiatric Clinics, 14(1): 163–176.

Weiler, I. J., Hawrylak, N. and Greenough, W. T. (1995). Morphogenesis in memory formation: Synaptic and cellular mechanisms. Behavioural Brain Research, 66(1-2): 1–6.

Wilson, M. (2002). Six views of embodied cognition. Psychonomic Bulletin & Review, 9(4): 625–636.

Wilson, A. D. and Golonka, S. (2013). Embodied cognition is not what you think it is. Frontiers in Psychology, 4: 58.

Zhang, F., Gradinaru, V., Adamantidis, A. R., Durand, R., Airan, R. D. et al. (2010). Optogenetic interrogation of neural circuits: Technology for probing mammalian brain structures. Nature Protocols, 5(3): 439–456.

Zhu, L., Aono, M., Kim, S. J. and Hara, M. (2013). Amoeba-based computing for traveling salesman problem: Long term correlations between spatially separated individual cells of Physarum polycephalum. Biosystems, 112(1): 1–10.

Index